# Field Guide For the Management of
# URBAN SPIDERS
## 2nd Edition

Stoy A. Hedges and Richard S. Vetter

EDITOR: DAN MORELAND
LAYOUT AND DESIGN: ANDREA VAGAS

PUBLISHED BY GIE MEDIA

ISBN: 978-1-883751-37-1
Library of Congress Control Number: 2012930156
Copyright 2012

All rights reserved. No part of this book may be reproduced or transmitted in any form or by any means, electronic or mechanical, including photocopying, recording or by any information storage or retrieval system without permission in writing from the publisher.

Address all correspondence to: GIE Media, 4012 Bridge Avenue, Cleveland, OH 44113.

# ABOUT THE AUTHORS

STOY A. HEDGES began his career in entomology at the age of two when he would collect honey bees in jars and turn over rocks looking for bugs. He received formal education in Urban Entomology from Purdue University in 1981, graduating with a B.S. degree. Hedges began his technical services career at ChemLawn, serving as the Technical Manager for its PestFree division. He also served as Director of Technical Services — Pest Control for Waste Management Urban Services, Inc.

Hedges joined Terminix in 1990 as Manager, Technical Services, Pest Control, and currently serves as Director, Technical Services. He is responsible for developing training programs and pest management strategies and procedures involving all structural pests, including wood-destroying organisms.

Stoy Hedges

He provides technical support to all Terminix branch and franchise operations and corporate departments. He was instrumental in helping to develop ASPIRE, Terminix's comprehensive basic training program.

Hedges is a Board Certified Entomologist, a professional sanitarian, and a member of the Pi Chi Omega pest control fraternity. He is well known throughout the pest control industry as an innovative thinker and was instrumental in popularizing the use of vacuums in structural pest management.

He has authored more than 150 articles for Pest Control Technology and Service Technician magazines during the past 25 years. In February 2001, PCT published a compilation of his best articles, titled *ANThology, The Best of Stoy Hedges*.

Hedges has served as Editorial Director for the 8th, 9th, and 10th editions of the *Mallis Handbook of Pest Control*, which is the prime reference guide for the structural pest control industry. In addition to his field guide, he is also the author of four other PCT guides for pest management professionals: *PCT Field Guide for the Management of Structure-Infesting Ants* (now in its 3rd Edition), *PCT Field Guide for the Management of Structure-Infesting Flies*, and *PCT Field Guide for the Management of Structure-Infesting Beetles*, Volumes I & II.

Hedges was named PCT magazine's Professional of the Year in

1997, and in October, 2000, he received the PCT/Syngenta Crown Leadership Award. Also in 2000, Hedges was recognized as one of the 25 Most Influential People in the Pest Control Industry by PCT magazine.

He currently serves on the Purdue Department of Entomology Planning Committee and has done so since 2007. Hedges is a regular speaker at pest control conferences including 26 consecutive years at the Purdue Pest Control Conference. He also performs training seminars for Terminix operations countrywide for food processing facilities that are Terminix customers, and has made presentations to EPA, FDA, and Department of Defense personnel.

RICHARD S. VETTER was born and raised in the blue-collar community of Elmont, Long Island, New York, the home of horse racing's famous Belmont Race Track. He spent his childhood developing his scientific skills by going through the usual kid stages of studying rocks, insects, and dinosaurs. Other than that, much time was spent playing unorganized sports in the street in front of his family's home. However, there was always a special curiosity and affection for animals, trying to understand what they did and how and why they did it.

Richard Vetter

Whereas many kids of the era were protesting the Vietnam War or becoming active in the counterculture, Vetter avoided the foibles of teenagedom and followed the academic pathway, graduating near the top of his high school class. He then attended Adelphi University in nearby Garden City, New York, where he earned a Bachelor of Science degree in Biology in 3½ years, graduating summa cum laude. After Adelphi, graduate school beckoned and Arizona State University in Tempe was home for the next two years, ending in a Master of Science degree in Zoology. His thesis topic was defensive behavior of the black widow spider. This started a lifelong interest in spiders, especially spiders of medical importance.

After graduate school and through serependity, Vetter was alerted to and secured a staff research associate position in the Entomology Department at the University of California, Riverside. His first job was in a moth sex pheromone behavior lab; Vetter was predominantly a bug farmer, running the rearing operation. During this period, he dabbled in spiders but did not pursue active research. After 12 years, Vetter was transferred to a honey bee lab where he maintained 40 to 70 colonies of honey bees. It was through this exposure and, again, some serendipity that Vetter started on a path of publishing on venomous arthropods. About the same time, the Entomology Department's spider-knowledgable museum curator was retiring. Vetter's modest goal was to merely learn the names of the spiders so he could provide spider curatorial services at UCR.

However, an incident in 1992 changed this modest goal into one that would lead to eventual national prominence. A newspaper story emerged about a Southern California woman becoming a quadruple amputee, the injuries of which were blamed on the brown recluse spider. Vetter innocently started inquiring about the brown recluse and very quickly the answers came back that they were not in California. Conversations with arachnologists across the country mirrored the California hysteria: no recluse spiders but plenty of stories about bites and injuries. Because there was no arachnological expert to consult on this subject, quoted sources were full of errors, myths, hyperbole and flat-out lies. Vetter saw an opportunity to fill the niche as the brown recluse expert for the country by putting out solid scientific information about this spider and its alleged bites, as well as countering the lunacy about bites from a spider that is non-existent throughout much of North America. Thus, the crusade started.

After a decade with many of his publications in medical journals, there is now a very different tone in the medical literature regarding how spider bites should be diagnosed. Besides the brown recluse issues, Vetter also addressed myths about yellow sac and hobo spiders, an Internet spider hoax, and the psychological basis for peoples' persistent belief in spiders as the only cause of the little red dots on their skin.

Vetter is a member of arachnological societies in America, Great Britain, Japan, Germany, and Spain, as well as the International Society of Arachnology. He has given about 200 seminars throughout the country and was an invited symposium speaker in Canada and Brazil. As of 2011, he has 117 publications in scientific and medical journals, three book chapters and does not plan to stop any time soon. He has given two spiders their scientific names and has had three arachnids named in his honor.

Vetter is finishing off his University of California Riverside career working in the lab of well-known urban entomologist Mike Rust, where he is working on control of spiders, ants, yellowjackets, fleas and termites. As of the assembling of this book, after 31 years of service at UC Riverside, Vetter may be looking at an early retirement in 2012, however, this will allow him to pursue many spider projects that have been on backburners for years. The next decade should be very productive for his arachnological research (until he spends his last years sitting on park benches feeding pigeons) and he plans to maintain his connections to the pest control industry, particularly as it relates to spider issues.

# ACKNOWLEDGMENTS

I would like to acknowledge the late Dr. Mark S. Lacey who encouraged my interest in more closely studying spiders in the early 1990s and which led to the publication of the 1st Edition of this field guide in 1995. Mark was a good friend and teacher, and his presence in our industry is sorely missed. Rick Vetter's expertise and patient counsel have continued where Mark left off in my understanding of things arachnological. And my wife Kathryn's patience with spiders in the house is also greatly appreciated. — *Stoy Hedges*

Almost all arachnologists are self-taught because there are few venues in which to learn this material. Therefore, I would like to thank those seasoned arachnologists who took the time to help a struggling novice as I navigated the tangled web of spider taxonomy, particularly Norman Platnick, the late Vince Roth, Charles Griswold, and H. Don Cameron. My hope is that this book will be a successful effort to pay it forward and teach others what I have learned in the last two decades. — *Richard Vetter*

# DEDICATIONS

This book is dedicated to Dr. Mark S. Lacey whose love of spiders helped educate hundreds, if not thousands, during his long career in the pest control industry. His life may have been cut short, but the impact of his passion for eight-legged creatures continues in our industry today.

# TABLE OF CONTENTS

About the Authors ............................................................... 3
Acknowledgments and Dedications ................................... 7
Preface ................................................................................ 11
Introduction ....................................................................... 15
Using This Field Guide ..................................................... 19
Basic Spider Biology ........................................................ 21
Spiders and Human Society ............................................. 33
Health Aspects of Spiders ................................................ 39
Care and Maintenance of Spiders in Captivity ............... 57
Basic Spider Anatomy ...................................................... 65
Control Techniques ........................................................... 77
Identifying Spiders ............................................................ 91
Mygalomorph Spiders ..................................................... 103
Tarantulas ......................................................................... 105
Other Mygalomorphs ...................................................... 109
True Spiders ..................................................................... 113
Imported or Banana Spiders ........................................... 115
Crevice Weaver Spiders .................................................. 119
Jumping Spiders .............................................................. 123
Wolf Spiders .................................................................... 129
Nurseryweb and Fishing Spiders ................................... 135
Spitting Spiders ............................................................... 139
Ground Spiders ................................................................ 143
Sac Spiders ...................................................................... 147
Lynx Spiders .................................................................... 153
Woodlouse Spider ........................................................... 157
Crab Spiders .................................................................... 161
Recluse or Violin Spiders ............................................... 165
Orb-Weaving Spiders ...................................................... 183
Cellar Spiders .................................................................. 191
Funnel Weaving Spiders ................................................. 197
Comb-Footed Spiders Excluding Widows .................... 205
Widow Spiders ................................................................ 211
Meshweb and Flatmesh Weaving Spiders ..................... 221
Relatives of Spiders ........................................................ 225
Glossary ........................................................................... 233
References ........................................................................ 237
Illustration/Photo Credits ............................................... 243

# PREFACE

Spiders are almost everywhere, but, if not, they may soon be by ballooning in! Yes, this is one method by which spiderlings (young spiders) and small adult spiders are able to use the wind and something that spiders produce (silk) to their advantage and travel great distances. The 2nd edition of the PCT Spider Field Guide has been thoroughly updated, and Stoy Hedges has collaborated with Rick Vetter, a professional arachnologist, to bring current arachnological information and up-to-date use of spider taxonomy (Platnick, N. I. 2011. The World Spider Catalog, version 12.0. American Museum of Natural History, online at http://research.amnh.org/iz/spiders/catalog. DOI: 10.5531/db.iz.0001), to you, the pest management professional. I've known Rick for over 20 years (initially during an American Arachnological Society meeting), and he is well versed in spider biology and taxonomy as well as venoms and medically important arthropods. Spiders are highly important predators in ecosystems and reduce insect numbers, often pest species, existing in and around human habitations. There is also the association of envenomation and spiders, and this field guide will explore the myths and realities of spider bites from a trained professional's view.

Louis Sorkin

Rick lends his arachnology expertise to this 2nd Edition, greatly expanding basic spider biology and taxonomy over the previous edition. You will learn about spiders and their natural history, ways to identify certain species or, at least, be able to identify to family certain unknown spiders that come your way. A taxonomic key is provided for identification purposes plus important information regarding particular species is also included. You will also learn that not all spiders produce webs, but, yes, they all produce silk. There are those specialized webs that are not the typical orb web style or cobweb that you may come across or your clients will have at their homes or businesses, but may be elaborate silken sheets, funnels, tubes, or strands that pass unnoticed and escape detection but are clues, nevertheless, for correct identification. There have been recent cases of spider webbing left on trees as millions of spiders climbed onto these after great floods. The webs lowered numbers of mosquitoes that normally would transmit malaria to humans. There are also other stories involving millions of tiny spiders who took up residence in a field and covered 60 acres with silk or thousands of spiders (a few species) that are normally not communal who produced thousands of webs and covered many trees. These are all rare natural phenomena.

In normal situations the professional is able to use these natural sticky traps to their advantage and examine the collections of arthropods in order to understand what may be pests in their client's homes or businesses. You will gain knowledge about spiders and be able to teach your clients about these fascinating creatures. They may ultimately decide that it is more interesting to watch them rather than remove them, especially if they are outside in webs or just crawling around and not inside their homes. Remember, they are providing an important service when alive.

The information contained in the field guide will enhance your spider knowledge. How many eyes does a spider have: 8, 6, 4, 2, or 0? What is the arrangement to the eye pattern? All information needed to help you identify a specimen provided by your client is covered here. Spinneret number and morphology, chelicerae and pedipalps: important items to know when trying to identify spiders. Spiders have eight legs, but there are modifications to their size, stature, and length, with claws, tufts, and setae plus other structures called scopulae and epigyna (epigynum is singular) to know about. Luckily, a glossary is included! So many things you have to know in order to identify spiders are covered in this field guide. The figures and photographs of commonly encountered species are included to aid you, too. Spider relatives including scorpions, sun spiders, whip scorpions, pseudoscorpions, whip spiders are also covered; all very interesting organisms not often seen by the general public but often feared when encountered. Public education is one important piece of the puzzle in pest management.

Stoy has authored and co-authored the PCT Field Guides to beetles, flies, and ants and in doing so has provided important information that can be carried to job sites and used on an everyday basis by the pest professional. In following these examples, he continues to assist in basic education and understanding of arthropods. Clearly knowing the proper identification of any pest species is key to understanding its biology and ultimately its management or control. In the present guide, he also handles pest management techniques for spiders when certain species have achieved pest status. Stoy, whose background is in pest management, provides techniques and information for those occasions where spiders have to be controlled. On many occasions, environmental modification can be used while at other times pesticides are necessary to achieve control.

Use of this publication will increase your knowledge of spider identification, their biology and also management techniques. This field guide, in addition to the others that he has provided to the pest management community, is a welcome addition to that special shelf in your library and also in your pocket when you as a pest management professional make house calls.

*Louis N. Sorkin, B.C.E.*
*Arachnologist, Entomologist*
*The American Museum of Natural History*

# INTRODUCTION

Spiders occupy an interesting niche in urban human society. They are one of the few groups of animals we run into on a regular basis no matter how far removed we are from a natural setting. They are also one of the groups of animals that we are taught to recognize as children and to which human personalities, either good or bad, are assigned; they appear in nursery rhymes, songs, and books (e.g., Little Miss Muffet, Itsy Bitsy Spider, *Charlotte's Web*). When was the last time you heard a children's story or song based upon a cockroach, termite, or tick? As we grow older, spiders are used as transgressors of evil or aggression, especially in Hollywood's movies. So spiders are easily identifiable by the average homeowner.

Despite, or maybe because of, these influences, spiders are one of the most common creatures for which pest control is requested. This is somewhat intriguing because spiders do not really decrease the standard of living for the average homeowner. They don't cause structural damage to a living space as do termites, carpenter ants, or rodents. They don't ruin lawns like gophers or ground squirrels. They aren't vectors of disease as are mosquitoes, ticks, and rodents. They aren't the source of allergic respiratory material like the feces of cockroaches. They are more of an aesthetic problem in that people see their webs as messy and a sign of poor housekeeping skills. They also create a psychological problem because the typical opinion toward spiders ranges from mild dislike to disgust to severe arachnophobia. Arachnophobia is a particularly interesting reaction. Think about it. A person outweighs a spider by 100,000 times, yet, the little eight-legged creature can cause restrictive fear in the great beast. It would be equivalent to you coming out of a Tokyo hotel, having Godzilla take one look at you then run screaming back into the ocean. Yet spiders have this commanding position over many of the population to the point that some people can't even look at a picture of a spider without having

Figure 1. Spiders are beneficial creatures helping to keep insect populations in check.

an adverse reaction.

Spiders play an important role in nature and the health and function of most ecosystems. As a top predator of the invertebrate world, spiders help keep insect populations in check (Figure 1, page 15). In fact, larger numbers of spiders in or around a home are an indication of a probable larger insect problem.

Spiders are arthropods, which means they have jointed legs. Under this broad umbrella of classification, they are related to insects, and crustaceans like lobsters and crabs. Within the arthropods, spiders are further segregated as arachnids. Being an arachnid means that it has eight legs when mature and mouthparts called chelicerae. Other groups of arachnids (the categories are actually called Orders) include scorpions, ticks and mites, harvestmen, vinegaroons, and whip scorpions (see the chapter on *Relatives of Spiders* for a more extensive discussion). In the order of spiders (Araneae), 41,000 species have been described worldwide within 109 families. In North America, there are about 3,700 species which belong to 569 genera and 68 families. Most of these are of academic interest only; they will never be found in homes or around structures. They can range from less than 1 mm long in body length (Figure 2) to about 11 inches in leg span for the large South American tarantulas.

Figure 2. Many spiders are very small. Here is the mature female (left) and mature male (right) of *Spirembolus erratus* on the back of a penny.

This book focuses on those families that the pest professional would have a high probability of encountering.

Two relatively unique evolutionary developments in spiders are attributed to their grand success at colonizing a wide variety of habitats, from tropics to high up on the Himalayas, deserts to swamps, and even underwater. The first is the development of venom for prey capture (Figure 3, page 17). Venom is a cocktail of proteins that can have a significant effect on its target organism. Mostly, venom is designed to immobilize the spider's prey. There may be venom components with differing functions. One venom component may be for immediate paralysis, which immobilizes the prey so it can't fly away or flail around

Figure 3. Venom is produced in elongate venom glands and delivered via a duct to the fang which injects it into its target. This is a dissected cephalothorax of a black widow.

with spines on its legs and injure the spider; usually this affects the nervous system.

However, in some spiders, the quick knockdown component is short-lived so there is another venom component that is slow-acting but long-lived or may cause death. Other components may be spreading factors that open up tissue inside the prey's body to allow the toxic components to have greater and faster effect. In rare cases, venom is used defensively as a protection, but most spiders don't attempt to bite a larger animal unless it is a last-ditch survival tactic. The second development that has aided spider survival is the silk glands. The emergence of silk has occurred in some insects, but it takes flight, so to speak, in spiders. Spiders use silk to line their burrows for buffering environmental effects such as humidity control and water repellency. They use silk to cover their eggs (Figure 4) and affix them in secure places. They spread the silk out in triplines from a burrow or a hiding place in order to catch prey. By making a web that is many more times its own body size, a spider can filter a relatively very large portion of the environment. The vibrations from an insect trapped in the web tell the spider much about its potential meal including size and location within the web.

Spiders are one of the first colonizers of new terrain. Because of their behaviors, there will be a continual onslaught of spiderlings landing on your clients' homes, and setting up shop to make a living. It is difficult if not impossible to live in a spider-free area. Many spiders disperse by ballooning when they are very small, where, on warm days with uplifting currents of air, they climb to the top of fence posts and vegetation, stick their abdomens skyward, and emit a strand of silk (Figure 5, page 18). At some point, the updraft of air is greater than the force

Figure 4: Spiders use silk for a variety of purposes such as prey capture, prey wrapping, and egg sac construction. Here, a spider guards two egg sacs.

Figure 5: Spiderlings disperse by ballooning where they exude a long strand of silk that catches the wind and carries them off to new destinations.

of gravity and it lifts the spiderlings from their earthbound locations and carries them into the air such that they become aerial plankton. Most spiders only travel a few feet or so, but some ballooning spiders have been collected at 15,000 feet with specially equipped airplanes and others have landed on ships that were miles from shore or any islands. During massive ballooning events, as millions of little spiderlings land, fields and fence posts can become completely covered with a sheet of silk. Ballooning is the prime reason that spiders are one of the first colonizers of new habitat after a volcanic eruption, decimation by other natural disaster, or merely the leveling of natural habitat and building of new homes.

Another way that spiders can get into homes is that they are very adept at squeezing through the smallest cracks under doorsills or around the edges of windows. Even a crack ⅛ inch in size is opening enough for a spider to slip on through. It would be virtually impossible to secure a home sufficiently enough to prevent spider entry.

# USING THIS FIELD GUIDE

As with any pest of concern, accurate identification is key in knowing how to deal with it. Especially now with the Internet where paranoid homeowners will spend hours trying to identify the pest in their homes and, typically, they will force an identification onto their creature no matter how inaccurate rather than to end the matter empty handed.

The key in this book was designed with the pest management professional in mind, which means the key was made so it is functional for someone who does not have vast experience in arachnology or even entomology, nor have the luxury to spend many hours poring over obscure papers to identify spiders. The reader also should be aware that this book was written to be useful to as large a portion of the United States as possible. Therefore, readers in special ecological areas such as Florida or Southern California may have trouble identifying some species which are very limited in range or which are very unique species found nowhere else in the United States. A balancing act had to occur with the chapters included in this book. We tried to cover as many families of spiders that would be common in urban homes but not include so many that it became an unwieldy mess.

# BASIC SPIDER BIOLOGY

Spiders are fascinating creatures that evoke wonder and awe as well as disgust and fear. Few arthropods can claim that kind of following. As with most animals, once you start delving into them, you start to see a whole new world through their eyes, and even more so with spiders because they have eight eyes and you only have two. The North American species are strictly predatory (although some might siphon nectar occasionally from special plants), they exhibit a tremendous variation in the ways that they have evolved to sustain life, and have developed special traits to ensure their survival.

### Life Cycle

*Egg sac.* A spider's life starts off as an egg. Usually, many eggs are laid within a protective silk egg sac (Figure 1). The egg develops and within about a week, depending upon temperature, primitive legs start to form on the outside of the egg surface (Figure 2). Within about two weeks, the egg hatches, and the spider enters the first stage of life called an instar. At this point, the 1st instar spiderlings are often pale in color and pretty much helpless as they can't move very well and often just lie there flailing their legs (Figure 3). They do not feed but instead live

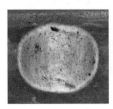

Figure 1: Most spiders produce a sac made of silk in which to house and protect their eggs.

Figure 2: As the spiderling develops, rudimentary legs are visible on the side of the egg.

Figure 3: First instars are mostly helpless, are not capable of moving around, and spend this stage inside the egg sac. They molt again before they emerge. The two crumpled white objects in the picture are the remnants of the egg.

off the nutrients of their egg yolk.

Within another week or so, they molt again and enter the 2nd instar where they now may have much more pigment in the body, sometimes even looking similar enough to the adult coloration that they are recognizable to genus or species. The second instar spiderlings stay inside the egg sac for another week or so. After about a total of a month from egg to 2nd instar, they are ready to emerge from the silk egg sac.

In many species, the spiderlings chew their way through the egg sac wall at which point, spiderlings come pouring out of the sac like clowns from a small car at the circus (Figure 4).

Figure 4: Second instar spiderlings have cut a hole at the bottom of the egg sac and have emerged.

However, in some species, the mother has to tear a hole in the sac to help her babies escape. In web spinners, the babies often hang around the web for a while before dispersing. This period could last a week for something like black widows (Figure 5) or more than a year for young tarantulas sharing a burrow with their mother. Hunting spiders may hang around the egg sac for a while, but in wolf spiders, the babies climb on mom's abdomen and get carted around on their arachnid taxi.

Figure 5: Black widow spiderlings hang around in their mother's web for about a week after emerging from the egg sac.

The number of eggs per egg sac varies widely in spiders. In some very small tropical spiders, the mother lays one egg which weighs 40% of her body mass. The spiderling emerges and is about half the size of the mother. Most hunting spiders, such as the brown recluse, lay about 20 to 50 eggs per sac. Tarantulas can lay several hundred eggs. Web spinners lay more with the black widow producing about 200 per egg sac and orbweavers up to 2,000 eggs.

Many species of spiders exhibit maternal care. For most, this just means living in a protective silk retreat along with the egg sac in order to guard it against predators and parasites. Other spider species, like the black widow, take it a step further and move the egg sac around from the retreat and place it out in the sunshine so it will warm up, accelerating spiderling development. A South American spider that is common in the southeastern United States and in Southern California actually will leave food for her emerged spiders and, in one case, ripped a fly apart, so the head, thorax, and abdomen were separated after which she took one of the parts back to her retreat and left the rest for her offspring to feed on. In a few spiders found in other parts of the world, the mother actually offers herself to her babies for their first meal.

*Spiderling dispersal.* Spiderlings disperse by a variety of mechanisms. Many just head out on their own from the mother's lair when they are ready to seek their fortunes in the new world. Wolf spider babies, who are clinging to their mother's abdomen, just fall off here and there until none are left. Many small spiders use a behavior called ballooning which was already covered in the introduction.

*Molting.* Because a spider has a rigid exoskeleton, the legs and cephalothorax cannot grow during an instar although the abdomen can plump up nicely. However, in order to grow larger, the spider must shed its old exoskeleton. When a spider gets to a certain phase of a particular instar,

Figure 6: In order to grow, spiders must molt, emerging from a smaller exoskeleton. This spider is just starting to emerge from the old skin.

Figure 7: As the spider molts, it has to remove its eight long legs from the old skin. This is a risky task; sometimes the spider becomes stuck and dies.

Figure 8: The spider is almost completely free of the old exoskeleton.

it starts to build a new exoskeleton underneath the old one. The spider stops feeding at this point, for about a week prior to molting. The legs darken as a sign of newly formed hairs. The spider takes up a position in its web or in its silk retreat. The spider contracts its abdomen, which pushes internal fluid up into the front part of the body. The carapace cracks and falls away as it does on the abdomen. The pressure pushes the cephalothorax up and away from the old exoskeleton and the spider starts the tedious and potentially dangerous process of trying to extract its legs from the old exoskeleton's leg tubes (Figure 6, page 23). Rhythmic pulsing of the body and continued pressure keeps the spider extracting itself (Figure 7, page 23) until the point where the legs are free from the old skin (Figure 8, page 23). Now, the spider is extremely soft and extremely vulnerable to damage or predation. It may hang for a few minutes while its new exoskeleton expands to its new size. The spider is rather quiet now while it starts to harden the new cuticle (Figure 9). Eventually, after sufficient hardening the spider is able to move more confidently around its environment. After a few days, the spider's fangs harden enough and feeding muscles are sufficiently strong that it will resume feeding. Molting is a very dangerous time with occasional spiders dying because they were not able to free themselves from their old skin.

Figure 9: After it molts, the spider is very soft and vulnerable to predation or damage. This pale brown recluse lacks the typical violin pattern which eventually will darken.

***Maturity.*** Spiders go through several molts before they mature. For the smaller spiders, it may only be two to four molts, whereas big spiders, such as the huntsman and fishing spiders, may go through 10 to 12 molts before reaching maturity. In females, there may be slight indication that she is nearing her final molt because a rudimentary bit of genitalia may be evident where the mature epigynum will form. However, most often there is little indication. Once the female molts, the sclerotized epigynum is present.

In males, in the stage just before the molt to maturity, the terminal segments of the pedipalps appear swollen (Figure 10, page 25). No features are present on the palp so it is useless for identification purposes.

However, this indicates that in the next molt, this spider will become a mature male. When this happens, the palpal bulb goes from featureless to developing numerous sclerotized structures that helps guide the inseminating piece into the female's genitalia pore for sperm transfer. In the male, the molt prior to maturity is referred to as the penultimate instar.

The common spiders molt every few weeks because most need to go from egg to adult within one growing season. Mygalomorph spiders may only molt once a year. For most spiders, the molt to maturity is the terminal molt; however, in some odd species (e.g., *Kukulcania*), they can be mature, lay a fertile egg sac, and then molt again. However, because the internal portions of the female genitalia are swept clean of sperm during the molt, she has to mate again before producing another egg sac.

Figure 10: The reproductive structure in the mature male is on the end of the pedipalps which are the dark structures on the front of the spider's body.

***Lifespan.*** Mygalomorph spiders such as tarantulas can live for many years. Typically, a male tarantula requires about seven to eight years before he attains maturity and then runs off looking for females to mate. Female tarantulas can live up to 20 years in captivity but probably do not live that long in nature. Many of the common spiders live their entire lives in one short summer season, emerging from the egg sac in spring, growing voraciously during the summer, mating in late summer, and then dying off in autumn with females leaving behind an egg sac for the next year. Recluse spiders in captivity are known to live four to 10 years and *Kukulcania* spiders also are long-lived. But most spiders have a lifespan of less than one year.

## Sexual Differentiation

As mentioned above, it is relatively easy to differentiate the mature males from mature females. The males have swollen palps which look something like boxing gloves and are carried around in front of the eyes. In many species, the males and females look very similar although a general rule for hunting spiders is that females will have larger abdomens to accommodate the pile of eggs she produces, and the males usually

Figure 11: The male (top) has a smaller abdomen and proportionately longer legs than the female (bottom).

have longer legs proportionately to the females. In other species, males and females differ in size, coloration or other features (Figure 11).

In web-spinning spiders, males are often times much smaller than the females in both body and leg length. In some orb-weavers such as the golden silk spider (*Nephila clavipes*), the males are so small that one might think they belong to a different species or maybe even be a parasite of the female.

In jumping spiders, which use their excellent vision and a visual courtship dance, there can be great sexual dimorphism where the female is usually decked out in drab browns and tans, but the male has brilliant colorations of cherry red, greens, iridescent blacks and blues, gaudy stripes and contrasting dots (see the pictures of the male and female jumping spider, *Habronattus pyrrithrix,* in the color image section of this book). Years ago, some jumping spider species were so different looking that early arachnologists placed corresponding males and females in different species and sometimes different genera.

## Courtship and Mating

Courtship in most spiders is initiated by pheromones laid down on silk lines, even in the hunting spiders (Figure 12). Males approach females and their courtship may consist of a variety of behaviors such as drumming with the palps or vibrating the abdomen which may have a pick rubbed against a series of ridges such that it creates a sound called stridulation. Using sensitive

Figure 12: The smaller male approaches the female using chemical odors (pheromones) which the female has released for species recognition and willingness to mate.

microphones, these vibrations can sound like an accelerating motorcycle engine. In spiders with excellent vision, the male may wave around his palps and legs, display colors, stridulate, or bounce his abdomen, among others. The female, of course, finds this ever so alluring. In crab spiders, the male ties down the female with a veil of silk but this is only ceremonial as the strands are flimsy and the female can easily break the silk once mating is over.

Once a female gives the signs that she is receptive, the male approaches and places his palp into her genitalia opening. This is a pretty slick trick because many of the spiders cannot see well at all, no sensory feedback is provided from the sclerotized structures of the palp because they lack nerves, and the male is often reaching under the female so insemination is pretty much done blindly. In green lynx spiders, the male makes a massive lunge from very far away; it is amazing that he is able to accurately hit the spot with his palp from such a distance. The coupling may happen face to face with the female raising up the front part of her body to allow the male access, or the male may climb on top of her and reach underneath. In web-spinners, the male may face in the same or opposite direction depending upon the species.

One of the great myths involving spiders is that the female eats the male after mating. This may happen occasionally in any spider species; however, the male would prefer to run off and try to mate with another female, if possible. For some species, the male does offer his body to the female as a light snack after mating; if she is hungry, she may accept, but if not, she may push his body away.

The more evolutionarily specialized spiders have two openings to the female's reproductive system (entelegyne). Sperm goes in one way and fertilized eggs come out the other. In these cases, it is of great benefit for a male to be the first one to mate with the female because he will then fertilize the majority of the eggs compared to any Johnny-come-latelys. Because of this, for some such as the widow spiders, it is not uncommon to find an immature female with several mature males in her web waiting for her to mature. (Those of you with teenage daughters should be able to relate.) Once the widow matures, a race by the males ensues to be the first male to copulate. It is also possible in some of these spiders that the female is only able to mate while she is still soft within the few hours of post-molting. Another advantage to mating right after molting is that the female's fangs are not yet hardened and so the male is safe from being eaten by the female. In addition to this strategy, a common habit among the web-spinners is for the male to break off part of his mating structures inside the female's genital opening or to ooze out a secretion that plugs the entrance. Both of these are mechanisms to prevent subsequent males from mating and introducing competing sperm. Very often, entelegyne females have no need to mate again and can produce fertile eggs for the rest of their lives. Once mated, the female stops producing sex pheromones and the males no longer find her web attractive.

In comparison, haplogyne spiders like the recluses, have only one genital opening so sperm goes in and fertilized eggs come out the same opening. For haplogyne males, the strategy is the opposite: they want to be the last male to mate with a female. Therefore, haplogyne females are willing to mate throughout their lives.

### Egg Laying

After a female has been inseminated and the eggs are fertilized and ready for laying, the female prepares a protective silk sac where she hides the eggs. For ground-dweller spiders, the first activity may be the creation of a mat of silk woven onto a flat surface. The eggs are then extruded as a sticky mass from the

Figure 13: Spiders often cover their eggs with a protective coating of silk to prevent predation, parasitism, and desiccation.

female's body onto the mat. The female then moves over the egg mass weaving silk to cover the eggs, which are then fully covered. The sticky egg mass dries out and the eggs then become separated and roll around like little ball bearings inside the egg sac. When ripping open the tough silk of a black widow sac, one must be careful not to spew the eggs all over the place because they will bounce very nicely and roll away and under things. For web-spinners, the female starts the egg sac as a little cup facing downward in her web, extrudes the sticky egg mass into the cup, then spins the rest of the egg sac around the egg mass (Figure 13). The eggs eventually dry out and fall to the bottom of the sac, which is often round but can take quite a variety of forms, depending on the species.

Most spiders hide their egg sac in a retreat or inside a protective

Figure 14: Cellar spiders carry their eggs sac in their fangs.

retreat under bark or in a rolled leaf. Wolf spiders attach the white pillow of a sac to their spinnerets and drag the sac around. Nursery spiders carry the sac underneath their cephalothorax, holding the sac with fangs and pedipalps. The pholcid spiders weave very few strands around the eggs such that they are eas-

ily visible through the silk and then she carries the sac in her fangs (Figure 14).

### Food Acquisition

Almost all spiders are predatory, capturing their prey live usually using venom to paralyze and eventually kill their dinner. (Just for completeness, one North American family of spiders [Uloboridae] lacks venom glands so they just wrap prey and start eating, and one tropical spider specializes in eating specific nutrient-producing portions of a tree so it is vegetarian.) A great variety of methods are used by spiders to secure a meal, and these will be covered below.

A spider feeds by exuding enzymes over its food so digestion occurs outside the spider's body with the spider using its sucking stomach to siphon up the nutrients into its body. The web-spinners and other spiders with weak jaws leave a corpse that is very much recognizable except that it is empty on the inside. The strong-jaw hunting spiders such as the wolf spiders, huntsman spiders, and the tarantulas, mash their prey up into an unrecognizable blob which, when discarded, looks like a crumpled piece of dark paper with few identifiable body parts.

*Active Hunters.* The active hunters may have a retreat they call home and return to on a daily basis (Figure 15, page 30), but these spiders run around hoping by chance to run into something edible and of sufficient size that can be overpowered. These spiders can be nocturnal or diurnal and run over the surfaces of leaves, on the ground, or on buildings. They may have excellent night vision (wolf spiders), excellent day vision (jumping spiders) or they may have poor vision, using airborne

Figure 15: Many active hunters, like this yellow sac spider, spin silken retreats in which to spend their time while not hunting.

or ground vibrations detected by the fine hairs on their legs. They do not use silk to entangle their prey, instead pouncing on prey, often just holding it in their fangs until the venom takes effect. Examples of active hunters include wolf spiders, jumping spiders, sac spiders, ground spiders, spitting spiders, lynx spiders, and woodlouse spiders.

*Passive Hunters.* Passive hunters sit and wait in areas where prey is active, waiting for the right moment to seize a meal. They are also known as sit-and-wait predators. Favorite spots for these spiders include flower heads or near ant hills. Crab spiders (Figure 16) are excellent examples as they match the coloration of the flower so the prey does not notice their presence (see the color image section of the book). When the flying prey starts feeding on the flower, the crab spider inflicts a bite just behind the head, causing instant paralysis. Passive hunters include crab spiders, fishing spiders, and tarantulas.

*Web-Builders.* Web-builders likewise exhibit tremendous diversity. The orbweavers build the archetypal Halloween spider web that spans a large space and filters a great area compared to that of the spider's body. Yet some orbweavers have evolved to where they just swing around a sticky glob of silk when they detect the wing beats of an approaching moth. Webs can be vertical or horizontal, or they can be like a funnel where the spider sits in a hole at the entrance of a retreat. Other webs can be individual strands that emerge from a retreat hole and act as an alarm system where the spider rushes out at the first jiggling of a thread. Webs may be sticky from adhesion or merely entanglement

Figure 16: Some crab spiders lie in wait on flower heads, mimicking their color in order to camouflage themselves from their prey. This spider is yellow and photographed on a green leaf for better contrast.

of arthropod leg sections (Figure 17). The silk may function to capture and prevent escape of prey, to trip up the prey so it falls upon another portion of the web, or serve just as an early warning signal that dinner is out there. Web-builders include orb-weavers, funnel weavers, comb-foot spiders, sheetweb spiders, and meshweb spiders.

Figure 17: Spiders use a variety of web designs made of silk to capture their prey.

# SPIDERS AND HUMAN SOCIETY

Just as spiders weave webs, their influence is deeply woven into human culture around the world. They are objects of admiration and beauty as well as perceived to be sinister malevolents who skulk under the cloak of darkness, doing their evil. In western culture, spiders are often thought of in negative terms; however, elsewhere in the world, spiders are welcomed into homes as harbingers of positive social encounters and future fortune. In many countries, it is bad luck to kill a spider. Some societies have used spiders to predict the weather; this may actually have some validity because it doesn't make sense for spiders to maintain an orbweb if rain is approaching so if orbweavers take down their webs, this could be a reliable harbinger of an incoming storm. Finally, in India, spiders are sometimes tossed over a bride and groom as we would throw rice or bird seed. Try that in North America and see how far you get!

**Mythology.** Spiders are very much a part of the mythology of ancient and not-so ancient cultures (Figure 1). Everyone is probably familiar with the story of Arachne whose weavings were so beautiful that a spiteful goddess changed her into a spider so she could weave for eternity. The Navajo tribe in America gave great status to spiders as the creators of the world. It was Spider Woman living in a hole in the ground, who taught the Navajo women how to weave blankets. However, at the request of Spider Woman in light of this transferred knowledge, a hole is always supposed to be woven into a middle of the blanket in honor of their arachnid teacher. Africans have many stories about the very clever Anansi, who could change form from human to spider and used trickery to gain advantage over other larger animals in the jungle.

**Poetry, Literature, Film, Fashion, & Art.** Spiders are very much present in poetry and nursery rhymes. Children learn about spiders early in their lives.

Figure 1: Spiders have fascinated and frightened people throughout history such that they wove spiders into their cultures.

"Little Miss Muffet" may have had good reason to be chased away by a spider. She was a real person and the daughter of 16th century English entomologist Thomas Muffet (or Moffett) although conjecture endures that no connection exists between the entomologist and the nursery rhyme. During those years, it was common for spiders to be used as medicine and swallowed to cure ills. What child does not learn to sing and do the hand movements to "Itsy Bitsy Spider?" Although many of the images of spiders are negative, some are portrayed positively. Charlotte of *Charlotte's Web* is "some spider." Actually, the author E. B. White consulted one of the premier arachnologists of the time (James Emerton) and provided very accurate depictions of spider behavior in the story including the correct orientation of hanging with its head down in the web (so she can think better said Charlotte) and having her babies disperse by ballooning at the end of the book. She introduces herself as Charlotte A. Cavatica. *Araneus cavaticus* is an actual scientific name of an orb-weaver from the New England and Great Lakes states. More recent books for children include the Miss Spider series. Shakespeare makes use of spiders occasionally in the couplets of his plays. In Robert Frost's poem "Design," a spider plays a very metaphysical metaphor. And for song, how can one beat the Australian classic, "There's a Redback on the Toilet Seat" (Figure 2). One area where spiders are relatively lacking, however, is in classic paintings. In contrast, of rather stunning beauty is the use of spider web designs produced by the ski clothing company, Spyder (Figure 3, page 35), whose decorative spandex festoons our best American downhill skiers. The sports teams of the University of Richmond in Virginia are known as the Spiders and the defunct San Francisco Spiders was a minor league hockey team.

Figure 2: Redback spiders are the widow species found in Australia and are responsible for many serious bites.

In Italy during the 13th through 17th centuries, spiders gave way to a form of dance. It was believed that the bite of a spider was fatal unless the bite victim danced vigorously for days. This happened in the boot heel portion of Italy in the city of Tarantos. This city's name gave form to the dance (the tarantella) as well as the condition (tarantism). The spider that was blamed for the bite was a wolf spider, which they called a tarantula. (Note: the spiders that we now call tarantulas were named after the Old World wolf spider even though it was a very different looking creature. When early European colonizers and scientists came to the New World, they decided that the large New World mygalomorphs reminded them of the Old World wolf spiders, and, hence, the name "tarantula" stuck.) Musicians were hired for the purpose of providing dance music for the alleged bite victims, and the festivals were usually held at the end of harvest. To make things even more festive, it was also a time of wine consumption, exhibitionism in regard to clothing, and inhibition in regard to morality. The dancers would eventually collapse in a heap and be cured. In reality, this was probably just the pagan Bacchanalian festival that was being repressed by the newly-in-power Christians who sought to stop the paganism. The pagans were able to get around the repression by claiming that it was the bite of a spider that prompted the dancing and they would die if dancing was prohibited. In reality, a spider was probably involved, but it was the Mediterranean widow spider (*Latrodectus tredecimguttatus*) whose bite would have been more prominent after harvest because the widows are common in wheat crops. Additionally, there may have been some truth to the dancing as a cure because widow venom affects the muscles and is very painful. Actual *Latrodectus* bite victims move constantly to try to lessen the pain. So dancing for many hours may have actually been a valid medical remedy for the bite of the widow spider. Today, this history has left us the musical term, tarantella, which is a fast-paced piece in classical music parlance.

Figure 3: The black widow is the icon for a company that makes ski apparel.

In the American frontier, whiskey was used to counteract bites from tarantulas, which were thought to be dangerous. Whiskey became known as "tarantula juice." Not surprisingly, some scallywags would carry around tarantulas and proclaim to have been bitten in order to get free offerings of libation as a cure.

But of course, spiders are used to scare people because it is one of the most common things that people already fear. Movies like "Arachnophobia" and "Eight Legged Freaks" will continue to pander to those who enjoy having the bejeezus scared out of them. And, of course, Ron Weasley of Harry Potter fame is a well-known arachnophobe both on screen and in real life, a phobia which surfaces several times in the movies as well as Harry having to fight amazingly large spiders in the forest. Also from Harry Potter, there is Hagrid's tearful loss of Aragog, the spider he raised from an egg. Finally, it would be negligent not to mention Stan Lee's famous creation of the comic book hero, Spider-man, who taught us that with great power comes great responsibility. However, when he was developing the character, one of Stan's superiors told him not to base the superhero on a spider because "people hate spiders."

**Functionality.** Humans have used spiders for a variety of purposes to enhance their lives (Figure 4). Aboriginal Pacific Rim tribesmen bend a long stick into a circle at one end. When a large golden silk spider makes its web in the circle, it becomes an excellent fishing device for catching fish in streams because when submerged in water, the silk becomes invisible, and it is strong enough to snare fish. Patches of spider silk were carried into war in previous centuries to use as a bandage for wounds. The silk was relatively sterile and quickened the clotting process. In a primitive form of pest control in Africa, communal spiders numbering in the hundreds to thousands would cover a branch with silk; the native people would remove the branch and place it inside their homes such that it was near the apex of the roof so that the spiders would eat the disease and filth-spreading flies. In World War II, the extremely tough silk

Figure 4: Spider silk is an amazing material, known for its strength and elasticity.

of the black widow was stretched across glass to make crosshairs for gunsights.

In current society, scientists are using spider venoms as probes to investigate how nerve cells work because the venom affects the various channels that allow or prevent changes in cell polarity and, therefore, nerve function. By investigating this action, it could lead to insight on various neurological conditions such as Parkinson's and Alzheimer's disease. Another use for spider venom is to protect brain cells during periods where local and temporary blood flow is blocked. Another avenue of investigation involves the use of spider venom to address erectile dysfunction; the bites of the South America spiders of the genus *Phoneutria* cause priapism, typically in young boys. Venom is also being investigated to some degree for its insecticidal capabilities because it affects mostly insects with minor effect on mammals. This type of specificity may help in the development of insecticides that are highly toxic to insects but are basically harmless to people and their pets. Finally, silk scientists are very actively investigating the molecular nature of silk because of its ability to stretch great lengths and recover to its original structure. Spider silk is able to absorb a lot of energy, which is why when an insect first gets caught in an orbweaver's web, it doesn't ricochet and get bounced back out the way it came in like it would with a rubber band (Figure 5). Silk is incredibly strong for such a lightweight material. If scientists can ever unlock the secrets to making artificial spider silk, it might be useful in the development of better parachutes and bulletproof vests.

Figure 5: Orb webs can intercept flying insects without ricocheting them out of the web because they absorb the energy of the insect.

# HEALTH ASPECTS OF SPIDERS

One of the biggest reasons that a pest professional is called out to treat a home for spiders is the fear that there might be toxic evildoers skulking behind the cupboard doors, plotting their next attack on the human inhabitants of the house. This is especially true of protective parents with small children in the home. This chapter will cover the health aspects of spiders, mostly dealing with the effects of envenomation but also discussing psychological aspects such as arachnophobia.

**What Spiders Do Not Do.** Before launching into the large and important topic of spider bites, it might be best to describe some misconceptions (some of which are amazingly fanciful in their construction) that are held by the general public and, in some cases, by pest professionals who are not as familiar with spiders as they should be.

A spider typically only bites as a last ditch defensive response when it feels threatened (Figure 1). Spiders do not bite people to feed on them. Spiders do not feed on human blood. Spiders do not bite people to soften up their skin so they can lay eggs on them or under their skin to feed their babies. Spiders are not attracted to humans because of their body warmth. People do not swallow eight spiders in their sleep in their lifetimes. Spiders do not use humans in any way, shape, or form to sustain their lives other than indirectly using our buildings and gardens as places to make webs and hunt for insect prey.

Figure 1: Spiders rarely bite unless threatened. This tarantula is showing a threat display and its fangs to warn the owner it doesn't want to be handled.

**What is NOT a Spider Bite.** Probably more important than the paragraph above, information is presented here on what is not a spider bite.

- ***Multiple lesions on the body at the same time.*** Typically, spiders bite once when they are getting near-fatally crushed. If someone complains that they have multiple "spider bites" on their body, then the cause points to something else like a bacterial infection or flea bites. Spiders cannot survive multiple near fatal squashing repeatedly and still inflict bites. This is especially true if the lesions are widely separated on the body (as in the ankle, buttocks, and neck); the spider does not run around the body inflicting bites all over the place. Multiple lesions are also NOT the work of several spiders working together. The chance of getting one spider bite in a lifetime is rare; the chance of getting multiple spider bites at the same time is statistically improbable. Also, a spider bite erupts at the site of the bite. One bite does not cause multiple skin eruptions away from the bite site.
- ***Multiple lesions over several months.*** Again, it is statistically improbable to acquire more than one spider bite in a short period of time despite what people want to claim. If someone is being afflicted over several weeks or months, once again, it is much more likely to be a bacterial infection or some blood-feeding insect, such as fleas or bed bugs, as the cause of the problem.
- ***Multiple people having lesions at the same time.*** People want to believe that nefarious spiders are lurking in the dark recesses of their house, waiting for them to go to bed so they can silently carry on their assault. More that one person having lesions at the same time reflects the classic contagious bacterial infection scenario or a blood-feeding insect situation.

**Bites.** Although spiders are considered by the general public to be a common source of skin lesions, in reality, rarely is a spider actually involved. Over 40,000 spider species occur in the world and only a few are known to cause significant medical incidents in humans. In North America, only the widow and recluse genera are proven to cause severe medical incidents.

Figure 2: It wasn't until the 20th century that the black widow's ability to cause toxicity in humans was confirmed.

**Black Widow Bites.** In the late 19th century, in North America, entomologists and medical personnel doubted that a little round black spider could be the cause of such intense suffering and occasional death in humans (Figure 2, page 40). As the 20th century progressed, more evidence accumulated that indeed the black widow was a spider that was toxic to humans. Great hyperbole and paranoia of widows developed in the early part of the 20th century but dissipated in the latter half as antivenin was developed and the effects of its envenomation could be lessened or eliminated. The medical syndrome for black widow bites is called latrodectism, after the spider's genus name *Latrodectus*.

*Bite Symptoms*—The venom of the black widow affects the nerve-muscle interaction. In the normal body, a muscle contracts when a nerve impulse is carried along the length of the nerve, very similar to the way electricity flows through a wire. When it gets to the muscle, chemicals (neurotransmitters) at the end of the nerve are released into a small gap between the nerve and the muscle, which quickly migrate across the gap, attach to receptors, and cause the muscle to contract. Then the nerve resorbs the neurotransmitters so it can release them again when needed. It is similar to flicking on a light switch (the muscle contracting) and then quickly switching it off again (the muscle relaxes). What black widow venom does is cause the nerve to release neurotransmitters into the gap and does not allow the neurotransmitter to be resorbed. So the light switch gets turned on, and it is prevented from being turned off.

So what happens in a black widow bite is that muscles are constantly being contracted without relief. This results in a variety of symptoms such as rigid stomach muscles, which can be mistaken for appendicitis, muscular pain throughout the body, and aches. Bite victims often describe it being similar to a bad case of the flu. Because of the pain, bite victims are very restless and may rock incessantly in their bed trying to reduce the pain. The venom also affects the nervous system that controls some of our automatic responses such that if one were bitten on the right thumb, sometimes the right arm will sweat profusely; however, the rest of the body will be normal. At the bite site, there may be a little red spot and possibly red streaks running away from the bite toward the central portion of the body (this is inflammation of the lymph nodes). However, there will not be massive tissue necrosis (e.g., rotting) as found in some brown recluse bites.

*Treatment*—In some of the older literature, recommendations included muscle relaxants and hot baths, but these have recently been shown to have little value. The current remedy for black widow bites is to use pain relievers, either opiate or non-opiates, or benzodiazepines, calcium, and magnesium, although there is some controversy over which treatment provides the best relief.

Antivenin works wonderfully such that bite victims can return to normal within 20 minutes or so. Even when a widow spider was found in the diaper of a howling baby, once antivenin was administered to the child within half an hour, the child was quiet and peaceful again. Antivenin also works if given after lengthy delays. One widow victim was bitten in the Australian Outback and took three days to reach medical attention. However, once given antivenin, the symptoms quickly faded. American physicians are somewhat hesitate to give antivenin because it is based on horse serum proteins, thus creating a risk of anaphylactic shock and death due to the antivenin. However, Australian physicians use it more frequently because the pain reduction is quick and effective. As long as the patient is retained at the hospital and observed for signs of anaphylaxis, the antivenin is very useful.

**Brown Widow Bites.** A short note should be included here about brown widow spider bites (Figure 3). In general, they are not nearly as intense as that of the black widow. In a study in Africa of their black widow species and the brown widow (the same species that is in the U.S.), the African black widow bites developed many of the same symptoms as the typical latrodectism seen in North America: intense pain, sweating, irritability. However, in regard to the 15 brown widow bites, the two most common symptoms were that it hurt during fang penetration and it left a red mark. Similar anecdotal descriptions have been offered for American brown widow bites. The spider has been in Florida for decades, and Florida's most prominent arachnologist states that he can only remember a handful of bites by this spider. In part, although brown widows can be incredibly common, they also roll up into a ball and fake death (Figure 4, page 43) readily so they don't put themselves in a position to bite very often.

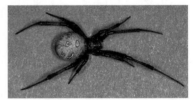

Figure 3: Brown widows are becoming very common in the southern United States but are much less toxic than black widows.

Figure 4: Brown widows are not especially prone to bite because they readily roll up into a ball and fake death.

Figure 5: False black widows look like black widows but are chocolate brown in color, not black, and never have a red marking on their belly.

**False Black Widow Bites.** The false black widow is in the same family as the black widow spiders but a different genus (*Steatoda*) (Figure 5). The false black widow is a common house spider where it is found. Its bites are described as being similar to mild black widow bites. Because these spiders are often found in homes, there is a greater chance of being bitten by this spider inside a home than by a black widow. The bite symptoms are mild, involve redness and soreness at the bite site, but resolve within a few days. However, because this spider is chocolate brown with a globular abdomen, it is easily mistaken for a black widow. Because almost no physicians are familiar with false black widows, even if presented with the spider, it would most likely be misidentified as a black widow. In one case, black widow antivenin was administered for a false black widow bite and the symptoms seemed to dissipate. One of the authors was bitten on the forearm by a male false black widow; a small, hard, raised mark was present about two weeks before disappearing.

**Recluse Spider Bites.** There are 100 species of recluse or *Loxosceles* spiders in the world, and every species tested so far has the enzyme responsible for causing skin lesions. Therefore, all recluse spiders should be considered medically important until proven otherwise (Figure 6, page 44). The medical condition caused by recluse spider bites is called loxoscelism.

Here are the four categories of loxoscelism:

**1. Unremarkable**—There may be a little swelling and a little redness but the wound heals on its own.

**2. Mild reaction**—Redness, swelling, itching, pain at the site but symptoms eventually disappear without scarring.

**3. Dermonecrosis**—This is the typical lesion that people think of when they think about loxoscelism, which involves the development of a nasty skin wound and the loss of tissue.

**4. Systemic**—These cases are very rare but involve the circulatory system with rupture of red blood cells and damage to the kidneys. Typically, these bite victims are small children, and the condition can be fatal.

Figure 6: The brown recluse spider, *Loxosceles reclusa*, is commonly blamed for a wide variety of skin lesions caused by other factors.

Although about 90% of loxoscelism cases in America end up with no significant damage (either unremarkable or mild reactions), it is the rare necrotic event that gets major publicity in both the general media and the medical literature. Think about it this way. Newspapers will report a massively destructive automobile crash on the highway because it attracts attention and people will read about it. However, would you read an article about someone denting a bumper in a parking lot? Of course not. But bumper denting accidents happen every day whereas horrific crashes do not and that is why the latter is news. Same thing applies to loxoscelism reports. If a brown recluse spider bites someone and nothing happens, no one reports it so no one is interested. No damage is the most common recluse bite manifestation. Nonetheless, the recluse spiders are still very much potentially dangerous, but luckily, horrific wounds are rare.

Figure 7: This woman was bitten by a male brown recluse 33 hours prior to the picture. She had swelling and a black eye but no necrosis. She healed without incident. By the smile, you should note that she is not in very severe pain.

*Bite Symptoms*—The bite of a recluse spider is usually not felt or if it is, it feels like pinprick or slight pinch. The venom causes the collapse of the blood capillaries so red blood cells cannot provide oxygen to the cells near the bite site. Within a few hours, the bite victim feels pain. As mentioned above, in most cases, the bite does not develop much beyond general symptoms of a generic spider bite: redness, swelling, and itching (Figure 7, page 44). However, in most severe bites, the lesion may start to turn purple, and the center of the wound develops a hardened crust after about a week. Eventually the crusted area separates from the perimeter of the wound and sloughs off. The remaining wound may take two or three months to heal.

*Treatment*—The old advice for treatment for loxoscelism in America was to cut out the affected area quickly in order to stop venom spread. However, this was later determined to cause more damage than good. Currently, it is felt that only in cases where damage looks like it will be extensive, that removal of skin tissue is advocated and then only after the wound stops spreading so that the healing process has already started. One Missouri dermatologist who specialized in brown recluse spider bites advocated leaving the lesions alone because many of them healed very nicely without medical intervention.

Many treatments are recommended for loxoscelism; however, there is no consensus on what works best as no proven remedies exist. The problem lies in that many recluse bites heal spontaneously. Without running a scientific study using a control group (that is, no medical treatment is given), doctors have no way to assess how well any particular treatment works. Because it is completely unethical to NOT treat bite victims, this information can never be determined from human bite victims. The best that can be done is to extrapolate from tests on animals, but animals also differ in their response to recluse venom compared to humans so the extrapolations may not always be valid.

Still, for most of the loxoscelism cases, simple first aid is sufficient. This is typically called RICE (Rest, Ice, Compression, and Elevation). However, because people would sometimes place ice directly on the skin, and cause even more damage from freezing the tissue, a better remedy is to just use cold packs.

In the cases of systemic loxoscelism, typically in small children, red blood cells are destroyed and hemoglobin flows freely in the blood stream. It gets filtered by the kidneys and passes through. In extreme cases, the child may pass dark urine having the same color as a cola soft drink. As drastic as this condition may be, immediate hydration and dialysis is typically sufficient to reverse the damage.

Antivenin has been developed for loxoscelism cases and is used extensively in South America. In America, because the *Loxosceles* spiders are not found nationwide and the fact that bites are not very common, the demand for pharmaceutical remedies is not sufficient enough to justify the expensive development of antivenin. Although antivenin works very well in bite victims, it seems that it needs to be administered within the first 24 hours of the bite to be effective. Because most bite victims do not seek medical attention until after 48 hours, an antivenin would not be effective. However, South American physicians feel that even if the antivenin is administered at a later date, it still speeds up the healing process and shortens hospital stays.

**Yellow Sac Spiders.** In the 1970s in the United States and in the 1980s in South Africa, yellow sac spiders (Figure 8) were implicated in studies as the possible cause of mildly necrotic skin lesions in these countries. In the U.S., the authors were cautious about implicating the spider, citing it as a possible source and reiterated that their proof was circumstantial. In South Africa, the author stated that the spider left characteristic symptoms of its envenomation, even though there was not one case of a verified envenomation in his reports. This information was broadly cited in medical publications to the point where it was well accepted that yellow sac spiders were medically toxic. However, looking at the evidence from each of these papers demonstrated that these authors were making conclusions that were not supported by their evidence.

Figure 8: Yellow sac spiders cause painful bites with mild symptoms. They do not appear to cause necrotic skin lesions as was previously thought.

In the U. S. paper, yellow sac spiders were incriminated because the researchers collected spiders in homes from people who had skin lesions with the result that the yellow sac spider was the most common spider in these homes. Pushing yellow sac spiders onto the skin of guinea pigs resulted in skin lesions; however, when they caused a bite in a volunteer grad student, nothing happened. Despite this lack of human reaction, the researchers stated that the yellow sac spider was a probable culprit in these wounds.

In South Africa, publications mentioned a stereotypic set of symptoms including the color of the venom once injected and a diagnostic distance between the fang marks. However, in a review paper later on, all of the alleged bites are listed as probable. Apparently, not one verified bite was listed in the lot.

Then in 2006, a paper with verified bites and a review of literature produced a strong counter-argument to the alleged involvement of yellow sac spiders in envenomation and disproved the notion that yellow sac spiders are common culprits in necrotic skin lesions. The paper described 20 verified yellow sac spider bites from the United States and Australia and reviewed 39 cases from the literature where bites were verified or probable (i.e., a spider was actually described in the incident in some way). In only one case was there a small pea-sized description of minor necrosis. All the remaining cases described a painful fang puncture with some swelling, redness, and itching for about two days, but the symptoms went away on their own. In two cases, there were episodes of recurring itching over several weeks. In all the other cases where a massive lesion was described, no spider was known from the incident. This evidence points away from a yellow sac spider as the cause of the lesion because their bites hurt like bee stings, even awakening sleeping victims when the bite occurred. It is highly improbable that one would be bitten by a yellow sac spider, feel nothing, and then develop a nasty lesion. Because of the pain of the bite, these people should have turned toward the source of the pain and found a yellow sac spider at the bite site. In addition, the South African papers state that the yellow sac spiders have a diagnostic bite distance of 6 to 8 mm between the fang marks. The yellow sac spider is only 11 to 14 mm long in body length. It is physically impossible for a spider this size with its normal-sized chelicerae and fangs to spread its mouthparts far enough apart and inflict such a bite. It would like you trying to do pushups by lying spread-eagle on the ground. Other evidence pointing away from yellow sac spiders is that analysis of their venom shows no enzymes known to cause skin lesions. Although it does have an enzyme that is capable of disrupting red blood cells, it is not capable of also causing skin lesions.

So what does happen in yellow sac spider bites? First of all, yellow sac spiders do bite and they do so readily. One report stated that the spider just crawled across someone's arm and without being squeezed (which is how most spider bites occur), the spider just bit down on the arm without provocation. The bite hurts in such a way that people have described it as a bee sting, a spring pushing through a sofa or a mattress, or a splinter stuck in their finger or toe. This pain is followed by mild swelling, redness, and itching. However, necrosis is not a symptom in verified yellow sac spider bites. Because toxicologists label a condition "rare" if it occurs three out of 100 times, the chance still exists in some cases that we have yet to see the full range of bite symptoms from this spider. But from what we have seen so far, no evidence exists to indicate that yellow sac spiders cause necrotic skin lesions. Therefore, this spider should not be included in the short lists of dermatological agents.

**Hobo Spiders.** In the late 1980s, the hobo spider, *Tegenaria agrestis* (Figure 9), was anointed as a spider newly associated with necrotic skin lesions in the Pacific Northwest. Before that time, lesions were, of course, blamed on the brown recluse spider despite the fact that there were no populations of brown recluses known to be living in the entire region. Word about the hobo spider as a new health threat was quickly spread by the news media and was unequivocally adopted by the medical community and the general public that this spider was the cause of mischief in the Pacific Northwest. At least one medical paper later claimed that the hobo spider was the most prevalent cause of skin lesions in the region. This spider was also labeled as the aggressive house spider possibly out of ignorance of associating its species name "*agrestis*" with aggression. The species name actually means "from the field" similar to agrarian and agriculture; the spider is not particularly aggressive.

Figure 9: In the late 1980s, the hobo spider was branded a cause of necrotic skin lesions, but new studies show this claim to be unlikely.

After hobo spider toxicity information was being widely spread for 15 years, in 2001 a venom toxicologist attempted to duplicate the original research that implicated the hobo spider. She could not induce skin lesions in the same strain of rabbits used in the original study. In 2004, a review of the medical literature showed that despite the fact that hobo spiders were listed in medical textbooks and many publications as being toxic to humans, the body of "information" that offered proof was extremely flimsy and, in some cases, contradictory.

Looking at the literature, only one verified hobo spider bite in a human is reported where a necrotic lesion developed. This involved a person who also had phlebitis, which can also cause skin lesions, and the person did not seek treatment until 10 weeks after the bite. The entire series of symptoms associated with hobo spider bite envenomation was based on circumstantial evidence. This same type of incrimination happened in South America with wolf spiders and in Australia with the white-tailed spider. It was later shown to be completely incorrect much to the embarrassment of the medical communities of each region. The alleged case histories of hobo spider envenomation in the medical literature did not involve verified bites, and if one were to carefully read the accounts, it was obvious from the description of the episode that it could not have been a hobo spider. That is "the victim walked into the hobo spider's web;" hobo spiders make their webs close to the ground so it is impossible for a person to walk into a hobo spider web. It seems like this might more likely be an orb-weaver, even though no spider was verified in the incident.

Additional general contradictory information is that the hobo spider is a non-native spider that became established in the Pacific Northwest around 1930. It is also not known to be medically important in Europe. Considering that the Europeans had several centuries of a headstart in gathering information on the environment and their relation to it compared to the United States, it is highly suspicious that this spider is not considered toxic there. Also, the hobo spider is a fairly large spider such that if it bit someone, the fang penetration alone should be of sufficient pain that one would take notice, probably find the spider at the scene of the envenomation, collect it, and produce it for identification. It seems highly suspicious that this spider has been in highly populated areas of the northwestern quadrant of the United States (Seattle, Portland, Salt Lake City) and British Columbia, in some places for more than 80 years and only one bite from this spider has been verified. Finally, several additional spiders of the genus *Tegenaria* occur in both Europe and the United States with none of them being known to be toxic to humans (Figure 10). Typically, members of the same genus share the same venom components and, therefore, they are usually all

Figure 10: The giant house spider is closely related to hobo spiders and is not considered toxic which indicates that the hobo spider may be wrongly blamed for bites.

considered medically important. For example, all recluse spiders tested to date have the necrotic enzyme sphingomyelinase that causes skin lesions, and almost all of the widow spiders have the venom component that causes latrodectism. It is quite suspicious that of all the *Tegenaria* spiders in the world, many of which are house spiders, that only the hobo spider is considered medically important. Finally, a study published in 2011 investigated the possibility that hobo spiders harbored bacteria that cause medical issues in humans including determining whether hobo spiders were able to pick up MRSA bacteria from infected petri dishes. Although it was not surprising that many bacteria were isolated from hobo spider body parts, none were virulent to humans. In addition, hobo spiders were not able to vector MRSA bacteria when exposed to it in petri dishes, further diminishing the likelihood that they are involved in skin lesions in humans.

For the moment, until a series of verified bites of hobo spiders can be presented to show the range of what happens in an envenomation, it is best to hang a question mark as to whether the hobo spider is or is not capable of causing necrotic skin lesions. Too many mistakes have been made in the past in regard to jumping too quickly to blame a spider bite as the cause of skin lesions. It would be better to await solid evidence so that mistakes are not made especially when the chances of misdiagnosing and mistreating something as a spider bite actually involves a much more dangerous non-spider medical condition that could be fatal. However, similar to other incidents in the world, the hobo spider may have been incorrectly elevated to the level of medical importance, and when the information is finally available, hobo spider bites will probably be similar to bites from the general spider population.

**Bites From Other Spiders.** Virtually any spider that is large enough and has cheliceral musculature strong enough can pierce human skin and inflict a bite. However, spider venom has evolved to paralyze and subdue prey, mostly insects, and this paralysis usually involves affecting the arthropod's nervous system. Because arthropod nervous systems use different chemicals for communication than do humans, almost all spider venoms will have little effect on us. The fact is a spider is trying to subdue something about its own size, not something that is 100,000 times bigger than itself. Really, if you were able to take down Godzilla as prey, what would you do with all the leftovers? But this does not mean that bites have no effect. The spiders are still injecting a foreign protein into your body, and the immune system will still perceive the alien venom as an intruder.

Almost all spider bites occur as a last ditch defensive response to being squashed. Those with strong cheliceral muscles will cause pain because their fangs are piercing your skin. However, most venoms that are injected have little effect on humans. The most common symptoms for a generic spider bite are redness at the bite site, possible swelling, and, as with most insect and arthropod bites, itching. The symptoms may persist for two or three days but will dissipate after that. Verified bites from other spiders include orb-weavers, funnelweb weavers, crevice weavers, wolf spiders, jumping spiders, lynx spiders (Figure 11), and comb-foot spiders, just to mention a few.

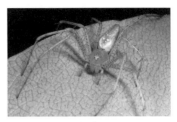

Figure 11: The green lynx spider has a mild bite, but it can also spit venom about 10 inches and has caused minor eye injuries.

**Conditions Misdiagnosed as Spider Bites.** One of the biggest advances in medical arachnology in the last decade has been the heightened awareness in both the medical and scientific communities, and even the general public, that many medical conditions will cause skin lesions and have been or could be mistaken for spider bites. Some of the medical conditions are far worse than any spider bite would be, and some are potentially fatal. When a physician diagnoses a "brown recluse bite" and treats it as such, the real condition is then still undiscovered and can progress unabated. Unless the physician can be convinced that the condition is not a spider bite, the patient may develop severe symptoms and possibly die. Some of the conditions mistaken for spider bites include:
- Infections by bacteria, fungus and viruses (including fatal group A *Streptococcus* and anthrax)
- Cancers, leukemia, basal cell carcinoma
- Diabetic ulcers
- Lyme disease
- Chemical or thermal burns
- Bed sores
- Reactions to blood medications

• Self-induced injuries from psychologically disturbed patients

One of the more disturbing aspects is that physicians in areas where brown recluse spiders do not live have historically diagnosed loxoscelism. This causes a variety of problems. First of all, the patients are being misdiagnosed, mistreated, and possibly could develop more dire symptoms and may die. Second, it spreads misinformation that a medically important spider is in the local area when it is not. Because this information comes from a physician, patients are very willing to accept this as the truth. A problem is then created in that the arachnologists and knowledgable pest professionals in the these areas know very well that recluse spiders are not found locally. This conflict gets further complicated because homeowners may start requesting pest control for brown recluses. Conflicts arise because the pest professional should avoid applying pesticide for a non-existent pest even if it is to pacify the client. But if a client has to sort out information between a physician and a pest professional, they are typically going to believe that their physician knows more about the medically important spiders in the area than a pest control professional. Using the maps in this book and several of the papers listed in the reference section may allow the pest professional to educate the client as to the inaccurate nature of their physician's diagnosis. On the positive aspect, a major shift has occurred in the medical literature where more papers on brown recluse bites are emphasizing that recluse bite diagnoses should only be made in areas of the country where brown recluses are found. Physicians are starting to get the message; however, it will take many more years before the correct information saturates their profession.

**MRSA (Methicillin-Resistant *Staphylococcus aureus*).** One of the biggest breakthroughs in recent medical arachnology is the discovery that bacteria resistant to antibiotics is one of the most common causes of skin and soft tissue lesions in humans and is frequently diagnosed as a spider bite. This infection is called MRSA (methicillin-resistant *Staphylococcus aureus*) (Figure 12, page 53). It is a contagious condition that is easily spread from person to person. It is found in places where people are housed in high density for long periods of time or where there is a high chance of contamination from body contact or inadvertent contamination. Such places include prisons, nursing homes, long-term health care facilities, military barracks, athletic locker rooms, and aboard ships. Humid locker rooms are breeding grounds for the bacteria, and the casual

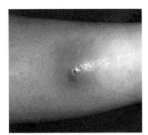

Figure 12: The bacterial infection MRSA is the most common cause of skin and soft tissue injury in the U.S. and is frequently mistaken for spider bites.

drying off and laying a towel down on someone's clothes could spread the infection. Football players have passed the infection on to the other teams during contact. Wrestling mats are now wiped down with alcohol to prevent spreading the infection.

Because these bacteria are resistant to many common broad-spectrum antibiotics, if it is treated like a typical infection, it will not be suppressed. At that point, MRSA can kill; some of its victims have been previously-healthy high school athletes, which is not a demographic group that we normally consider vulnerable to fatal diseases. It is estimated that 18,000 people a year may be dying from MRSA. If that is true, then this surpasses the annual number of deaths from AIDS.

It is important for the pest professional to understand a MRSA situation. For example, a nursing home may request pest control because many of their residents are getting "spider bites." Treatments may be applied but, of course, do nothing to stop the bacterial infection. Several weeks later, the client may request more treatment because more "spider bites" occur. At some point, the client may become irate and cancel service. The pest professional needs to understand what is and is not a spider bite (see Figure 12 above) and be able to communicate to a client in this type of situation that they may be dealing with a contagious bacterial infection, not a spider bite. Since information about MRSA has spread, on several occasions an arachnologist has been able to overturn a physician's diagnosis of spider bites as the cause of multiple lesions in a jail or a collegiate sports team and directed the concerned clients toward the correct diagnosis of a bacterial infection. If the knowledgable pest professional can do this, it should lead to greater confidence by the client and possibly an increase in business.

**Tarantulas and Urticating Hairs.** On the posterior portion of the abdomen of the tarantula, specialized hairs occur that are not more than 1 mm long; they are twisted or have barbs on them. If a tarantula feels threatened, it will quickly use its fourth pair of legs to rub this area and dislodge these hairs, which can become airborne. The barbed projectiles are

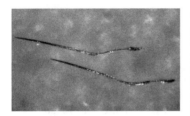

Figure 13: These are Type I urticating hairs from an *Aphonopelma* tarantula. They are about 0.5 mm in length.

called "urticating hairs" and can get imbedded in the nasal cavity, mouth, and eyes of would-be predators. The barbs make them difficult to remove, causing irritation to the unlucky recipient. In cases where urticating hairs have become imbedded in peoples' eyes, the only course of action is to let the body slowly digest the hairs, which may take up to nine months, a frustrating and time-consuming process.

North American tarantulas have Type I hairs, which are relatively harmless (Figure 13). Their function may be more to line the burrow such that if a mouse or other small predator attempts to enter the burrow, it encounters this irritating mat of hairs. Type I hairs are not very aerodynamic because these tarantulas spend most of their time in burrows but people still need to be aware of these risks. This includes wearing gloves and goggles when cleaning out a tarantula cage.

South American tarantulas are more arboreal and because they are exposed on tree trunks and such, their urticating hairs are more aerodynamic and, therefore, are more likely to get stuck in your nose and eyes. These are referred to as Type III hairs. But for the pest professional, this situation would only occur with an escaped pet or if this type of tarantula is kept as a demonstration animal, so it is of minor concern.

**Arachnophobia.** Spiders are extremely special arthropods in that they can invoke debilitating fear in humans (Figure 14). Although this may happen on occasion with other arthropods, one still does not get the full throttle, pedal-to-the-metal paranoia freakout dance that spiders can often elicit. In general, people have a negative view about spiders which ranges from mild aversion and disgust to full blown arachnophobia where people rearrange their lifestyle in order to eliminate or reduce the chance

Figure 14: Arachnophobes dislike spiders because they are hairy, they run fast and they have many long legs, whereas their toxic nature is low on the list of reasons for phobia.

of inadvertently encountering a spider during their day (or more with greater hyperventilation, at night when they are asleep and defenseless).

People with mild disgust of spiders also typically exhibit the same reaction to other creatures that are perceived to be dirty, vectors of diseases, or otherwise agents that reduce the quality of their lives. These other creatures include cockroaches, slugs, ticks, maggots, and bed bugs.

Arachnophobia is a very interesting and well-studied field. It often starts around the ages of 4 to 8 years (Figure 15). There may be a familial influence where a child learns the freakout dance from an arachnophobic parent or older sibling; however, it also can develop without this connection. Sometimes, there may be an association with a shock or fear incident that triggers the negative feelings. The niece of one of the authors turned on the light in her cabin at a soccer camp revealing a tarantula crawling across her pillow. From that point on, she has had a negative response to spiders. The aspects of spiders to which arachnophobes typically have the strongest reaction do not usually involve its toxic nature. Instead, it is the "spideriness" of spiders in that they have lots of legs, they have long legs, they are hairy, they run fast, and they show up unpredictably. Women are much more likely to be arachnophobes than men, although the lower numbers of men may also be due to men being less willing to admit to spider fear.

Figure 15: Fear of spiders usually starts at an early age and, in some cases, develops into a debilitating fear as an adult.

Arachnophobes can show up in any profession, even though the person understands that this is an irrational fear. When this occurs in the medical profession, it can lead to overemphasis on spider bites as a diagnosis for skin lesions because the physician already has a negative opinion about spiders. Quite amusingly, one of the authors has colleagues in his entomology department who were somewhat arachnophobic. These people worked with insects all day long for decades but just slap another pair of legs on the wee beastie and the creature is upsetting to them. One entomology graduate student grew up in Missouri and was indoctrinated to fear spiders because of the brown recluses in her area. Not only did live spiders make her run out of the room, but she couldn't stand to look at pictures of them in entomology textbooks. An entomology department secretary was so phobic of spiders that hearing the word "spider" in three consecutive sentences, she would physically shudder like she just took some bad tasting medicine. Of course, one of the authors (who shall remain nameless but you can figure it out), would hold his arms out wide and ask for a hug when he was wearing a t-shirt with a big spider on the front. Needless to say, he went hugless as was anticipated. There was even an account in a pest control magazine of a pest professional, new to the field, who had such severe arachnophobia that when a spider walked across his face while he was in a crawlspace under a house, he froze in fear and had to be extracted by the fire department who located him by the large amount of screaming coming from under the house.

Arachnophobia also can be initiated by a medical diagnosis, even when the probability of a spider being involved is highly unlikely. People who develop lesions and are diagnosed with spider bites can develop such a strong degree of arachnophobia that they need to go through extensive psychological desensitization therapy to alleviate their fears, which interferes with having a normal life. From northernmost Indiana (where they do not have brown recluse spiders), a man was diagnosed with a brown recluse bite and eventually passed away; his entire family developed arachnophobia, severe manifestation in two of the six members, which did not alleviate until spiders from their house were identified as harmless.

# CARE AND MAINTENANCE OF SPIDERS IN CAPTIVITY

Although most people would not consider keeping fuzzy and cuddly spiders as pets, some pest professionals do maintain live specimens as part of the educational portion of their business (Figure 1) or just for display purposes at work. Spiders are generally very low maintenance, do not require frequent feeding, and are typically silent unless they are crawling around in

Figure 1: Many pest companies keep spiders to use in educating children regarding the benefits of spiders and insects in our lives.

the cage and knocking things over. Below are some tips for the care and maintenance of spiders in captivity. General comments will be offered first, then specific spiders will be addressed, and other spiders will be reviewed toward the end of the chapter.

**General Care.** Let's start with an overview of this topic:
 • In general, spiders do not need water except those such as wolf and fishing spiders who live near streams or ponds or tropical creatures like the huntsman spider. Most spiders get sufficient fluid from their prey.
 • If a spider is kept in a small vial, just for transport or for overnight storage, place a piece of paper towel or a dry leaf inside the vial so the spider has something to crawl into or hold on to (Figure 2, page 58). If just thrown into a vial, the spider may not be able to crawl up the sides, at which point, it may just flail for hours, exhausting itself and dying.
 • Spiders DO NOT need air holes in a jar or container. There is plenty of air already inside the jar. As long as they don't starve to death, spiders can live for months inside a small container without suffocating.
 • In selecting food for spiders, typically something 70% to 100% of its body length is best. This will be a size that the spider normally will eat in nature and will be small enough to subdue and big enough to allow the spider to store a lot of nutrients so it doesn't need to feed again soon.

Figure 2: Placing a piece of paper in the container gives the spider a rough surface to attach silk. This recluse is almost obscured by silk she has laid down between the toweling.

- If possible, it is best to vary a spider's diet because certain prey may be deficient in specific nutrients. By varying the diet, it allows the spider to prevent nutritional problems. A visit to an outdoor light fixture at night during warm months can provide a wide variety of spider food choices.

- If spiders are immature and you are trying to get them to maturity as fast as possible, it is best to give them a break of a few days before offering more prey. Spiders that are overfed in captivity have a higher mortality than those fed moderately.

- Watch the spider's abdomen for plumpness as an indication of how often it should be fed (Figure 3). If very plump, feeding it might be fatal. If the abdomen is shriveled, the spider needs food, or possibly water, quickly. If the abdomen is shriveled and the spider is walking around in a very shaky, uncoordinated manner, it is dehydrated and may die soon if it doesn't get water immediately. Put a few drops of water in front of the spider, preferably on the tips of its legs so it can sense that water is available and watch if it starts drinking immediately. After a few minutes of slurping up water, it will have enough hydraulic fluid in its body to move normally again. Then it should be fed if possible.

- In choosing food, it might be best to offer prey that do not have chewing mouthparts. Moths and flies are great. Although most spiders will immediately attack prey, if the spider is getting ready to molt, it will not be eating and then after the molt it will be very soft and vulnerable. It is not uncommon to toss in a cricket or mealworm and come back the next day, the spider is gone and there is just one fat cricket in the vial. However, crickets and mealworms are commonly avail-

Figure 3: Abdomen plumpness is a good indicator of when to feed. The black widow on the left is underfed, the widow on the right is overfed, and in the middle is adequately fed.

**58** Care and Maintenance of Spiders in Captivity

able food from pet stores; just be careful that if a cricket is still alive the next day, remove it, watch the spider for a week to see if it molts, and then offer the cricket again later.

• If you see a spider in the act of molting or lightly pigmented from a recent molt, do not offer food for several days until its body hardens so it can actually eat.

**Black Widows.** The spectacular coloration of black widows rates them very high on the desirable list for display spiders. The fact that they are potentially deadly only enhances their celebrity. Widow spiders are one of the easiest spiders to maintain. They are usually satisfied to have a small space to make their web. An aquarium or cage that would fit a basketball would be more than enough room for a widow. She would be able to make her web and not realize that she is in a closed container. Placing a sturdy branch inside the cage provides a structure on which to make her web as well as being somewhat aesthetically pleasing. Although she would really like some rocks to hide under during the day in order to escape the light, this would make it difficult to see her. Affixing a clear plastic vial in the corner of the tank would give her some place secure to retreat to, from which she will make her web. It is better to place the vial at the bottom of the cage, otherwise she might make her retreat right under the lid which could make it difficult to open and close the container. Widows get all their fluid from prey so there is no need for a water dish in the container.

One very important caution is that widows can store sperm and produce fertile egg sacs for many months. If she lays an egg sac, there will be babies to deal with. If the lid to the cage is a typical screen, the second instar spiderlings will bolt right through the screen and be all over your office, which might take you out of consideration for "Employee of the Month." Watch for egg sacs. If one shows up, it still takes a month for it to hatch out so you have time to make a plan to deal with it.

Widow spiders live about two years but could live up to four years in captivity.

**Brown Recluses.** Maintaining a brown recluse in a container is possibly the ultimate in spider care because they are ultralow maintenance. They will want a retreat to hide in. You may want to tape a bent piece of cardboard to the side of the tank so that you can see the spider. Recluses can slide through the smallest of cracks so the lid needs to be very secure on the container. Recluses can't climb glass or plastic very well but might be able to get up to the top in the corners. If you keep a recluse in a small vial, it will need a piece of paper toweling around the perimeter of the vial. This allows the spider to lay down some silk which helps snare prey thrown into the vial and, if immature, allows the recluse something to grab on to so it can cleanly pull itself out of the old exoskeleton.

Recluses do not need to eat often. They are champions of food-deprivation, sometimes able to go six months without a meal. However, the males have smaller abdomens and, hence, need to be fed about once a month. Males also are more likely to eat if they have smaller prey than you would offer the female. Females seem to mature, eat a meal to get plump, and then just sit there waiting for Prince Charming to arrive and mate. She doesn't move around her container at night like the males do; therefore, females can survive with greater feeding intervals.

Female recluse spiders can live two to 10 years; males possibly two to three in captivity.

**Tarantulas.** With black widows and brown recluses, it is pretty easy to obtain one (just turn over some wood in a wood pile), it will only live a year or two before dying of old age, and if you get tired of it, you can always just toss it back into the woodpile. In strong contrast, the decision to obtain a tarantula is somewhat similar to obtaining a small pet: it will require a real commitment to care and can last 5 to 20 years depending on how old the tarantula is when you get it (Figure 4). And if you get tired of it, you can't just release into the wild, which would mean probable death because it wouldn't be able to dig a new burrow in time to secure itself.

There is a tremendous amount of information available for tarantula care as it is big business for the pet trade industry. We recommend the book *Tarantulas and Other Arachnids* by Sam Marshall for anyone interested in keeping tarantulas in captivity. The book is written for the amateur so it is devoid of a lot of jargon. This book provides excellent advice on all aspects of

Figure 4: Some species of tarantulas make good pets due to their gentle nature and being easy to care for.

Figure 5: Tarantulas molt while lying on their backs. Leave them alone! Righting them could cause legs to get stuck in the shed skin or possibly death.

tarantula ownership from things to look for when buying a tarantula as a pet to the pests of tarantulas. There is a lot to know about keeping a tarantula, and it cannot be covered sufficiently here in a few paragraphs.

Tarantulas are good spiders for keeping in captivity because, in nature, they typically settle into a burrow with their entire territory being the burrow and a foot or so around it. Therefore, they should not be roaming around the cage endlessly looking for a way to escape.

We strongly recommend contacting someone who has a lot of experience with tarantulas before you obtain one. Do not assume that the person in the pet store has sufficient knowledge about these creatures (although obviously, some pet store personnel are very knowledgable).

Critical things for tarantula care:

• Tarantulas molt while on their backs with their legs in the air (Figure 5). DO NOT try to turn the spider over because this will disrupt the molting process and could lead to an abnormal molt with legs stuck in the shed skin.

• Tarantulas have urticating hairs. If you choose a spider that is somewhat defensive and attempt to handle it, you may be subject to regular coatings with these hairs. Even if they don't dislodge hairs in your presence, hairs will probably be on the sides of the container and elsewhere. Always wear rubber gloves and goggles when cleaning out a tarantula cage. A paper face mask might also be very beneficial to prevent the intake of urticating hairs up the nose or in the mouth.

• Temperaments vary among the species. It will depend if you want something to handle occasionally (but be very aware of the urticating hairs) or just look at through the glass. Rose hair, pink toes, and Brazilian black tarantulas typically have a gentle nature as do a number of species from Central and South America. Novice tarantula owners should avoid Old World species (e.g., baboon tarantulas from Africa) as they are more defensive and may readily bite. Sites online that sell spiders are a good source of information regarding the temperament of different tarantula species.

• Tarantulas shed the urticating hairs on the back of their abdomen in regard to threats or other stress. These hairs replenish after a molt. However, if your tarantula often has a bald spot, then it is either being handled too much or it doesn't have a burrow where it feels sufficiently safe (Figure 6).

• If you handle tarantulas, it should be held over a soft surface no more than a few inches below your hand. Because of their big body and soft tissue, a fall from a few feet will surely be fatal. This also pertains to letting someone handle it and then they freak out and pull their hand away quickly, sending the tarantula flying to its death. Be very careful who you allow to hold your tarantula. It might also be wise to hold on to their arm if you are transferring the tarantula so they can't pull away fast and send the spider to a splatting death.

Figure 6: A bald spot on the abdomen of a tarantula could indicate stress and/or that it is being handled too much.

• Tarantulas are cannibalistic so only one to a cage.

• Tarantulas need a CONSTANT supply of water. Use a small shallow dish. Placing a small piece of sponge in the center of the dish allows prey to escape and not drown which would taint the water.

• Use a mister or atomizer to increase humidity in the cage. Humidity of 70% to 80% is recommended for tropical species, 40% to 50% for desert species. Temperature should be between 70° and 85°F. Above 90°F, they start to stress, below 60°F, they will not eat. Although desert species live in very dry conditions with greatly fluctuating temperatures and low humidity, tarantulas live in underground burrows where the conditions are very different than above ground and do not fluctuate much.

• Airflow is needed in the cage, more for humidity control than for the spider. Therefore, it needs a screen lid of some sort.

• The best substrates are vermiculite, peat moss, sphagnum moss, mulch, and potting soil. Tarantula experts differ on the opinions of what is best so you may need to seek out someone for help. DO NOT use soil from the backyard unless it is sterilized because it will be a source of mites and other pests that will threaten your tarantula's survival. DO

NOT use cedar shavings or similar material as these might grow mold and attract mites.

- If you wish to decorate the inside of the cage with living, low-light surviving plants, this will help regulate the humidity.

- Tarantulas crush their prey into a small ball to suck out the nutrients. They discard the bolus of prey at the end of the feeding. Remove the dead prey bolus because that will attract mites which feed on the remains.

- It would be wise to purchase foot long tweezers or a long-handled cooking spoon used for stirring food to remove dead prey and otherwise clean up the cage. You don't want the slight movement of your hand inside the cage to be mistaken for dinner. Although bites are not usually very toxic, it is better to be safe than sorry.

- Make sure you know what a male tarantula looks like. If you buy a mature male, it will not live very long.

- As mentioned in the tarantula chapter, don't bother trying to make a pet of a roaming male found in nature. He is just out looking for females for mating purposes and will be dead within a few weeks anyway. Just let him go on his way.

**Other Spiders.** Many other spiders can be desired to keep in captivity, and the variety is immense. Some people take orb-weavers and just toss them against the inside of a window where the spider makes a big web in the free space of the window frame. As long as it is fed regularly, it should stay in the same place. However, this is probably more than most people would be willing to tolerate. Most spiders will prefer a retreat in which they can hide. This could be a small cardboard tube or under a rock or piece of cardboard. It is a good idea to put something inside like cardboard or a crumpled paper towel so the spider has something to attach its web to in case the glass or plastic is too slick for web attachment.

The feeding regime will vary, depending on the spider. Jumping spiders run around during the day and will exhaust themselves. They need to be fed every week or so. Other spiders can go weeks or months between meals. Paying attention to the fullness of the abdomen should be a good indicator of how often a spider needs to be fed.

# BASIC SPIDER ANATOMY

This chapter will describe the various parts of a spider's anatomy focusing on those parts used for identification purposes. Almost all of the anatomy described here will be external as there is little that a pest professional would be using on the internal anatomy of a spider for identification. Some information about internal structures are mentioned, however, where it helps explain the external parts that are used for identification.

**The Cephalothorax Region.** Insects have three main body parts (head, thorax, abdomen). Spiders have two main body parts; the head and thorax are fused together and is called the cephalothorax; the second body part is still the abdomen. The cephalothorax is the head end of the spider and is the location for the eyes, it is where the legs attach, the mouthparts (the chelicerae and fangs) start the feeding process, and, internally, this is where the venom glands are located (in mygalomorphs in the chelicerae; in the common spiders in the cephalothorax itself). The top or dorsal surface of the cephalothorax is called the carapace. The bottom or ventral surface is called the sternum.

Figure 1: Jumping spiders are easy to identify by the large middle eyes in the anterior (bottom) row.

*Eyes*—The eyes of the spider are unique among arthropods in that they are more similar to human eyes rather than to insects, which have compound eyes made up of dozens to hundreds of facets. Whereas insects see and have to interpret a mosaic made up of many images, spiders that have good vision see more of an image like we do. However, many of the spiders have simple eyes more likely used merely to distinguish light from dark. Most spiders have eight eyes, some like the recluse and spitting spiders have six, in Europe some have four, two-eyed spiders run around the Southern California deserts, and cave spiders have no eyes. However, most of the spiders you will be trying to identify will have

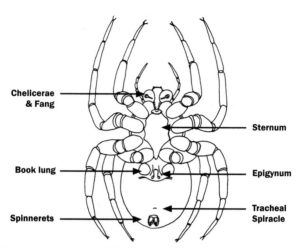

Spider anatomy – ventral view (after Bristowe)

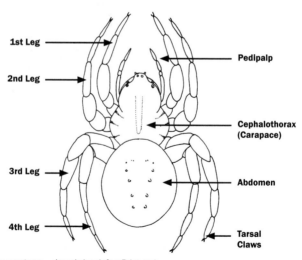

Spider anatomy – dorsal view (after Bristowe)

# SPIDERS VS. INSECTS

|  | SPIDERS | INSECTS |
|---|---|---|
| **Body Regions** | Two | Three |
| **Pairs of Legs** | Four | Three |
| **Wings** | None | None; one or two pairs in the adult stage |
| **Eyes** | usually 6 or 8 eyes | Can have compound eyes with 2 or 3 simple eyes |
| **Mouthparts** | One pair of chelicerae | One pair of mandibles; may be modified into various forms |
| **Pedipalps** | One pair | None |
| **Antennae** | None | One Pair |
| **Development** | Without metamorphosis; spiderlings closely resemble adults | Various development; without complete metamorphosis |
| **Food Digestion** | Occurs before swallowing | Usually takes place after swallowing |

eight eyes with a few having just six. Be forewarned, though, that very often a pair of eyes can be very small and very close together so they appear to be only one eye, which will throw off attempts to identify it.

The pattern of eyes is very useful in identifying some spiders to the family level. For example, only the family known as jumping spiders has a large set of anterior eyes in the middle of the front of the carapace with smaller eyes located behind (Figure 1, page 65). Jumping spider vision is so well developed that it rivals human eyes in the ability to distinguish objects. Wolf spiders have a row of four small eyes across the front but have two large "headlight" eyes above this row which allows for keen vision. With the well-separated eyes placed around their

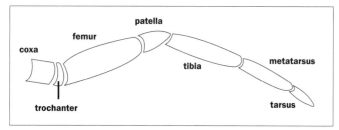

Figure 2: A spider leg consists of seven segments.

carapace, the wolf spider can see 360° around itself.

***Legs***—Without exception, all spiders have eight legs as long as they haven't lost any due to predatory attempts. Each leg is composed of seven segments (Figure 2). Nearest the body a small segment, called the coxa, connects the cephalothorax to the rest of the leg. The next segment, even smaller and called the trochanter, is usually only visible if you bend the rest of the leg away from the body. From here, the leg segment names become more familiar: femur, patella, tibia. The two end segments are the metatarsus and the tarsus. At the end of the tarsus are claws, which the spider uses for running or grabbing the web as it moves through the silk strands (Figure 3). A spider can have either two or three claws. If two claws, the spider usually also has tufts of hair at the end of the tarsus and/or along its ventral length. Two-clawed spiders usually are hunters that run around the ground or in vegetation. If three claws are present, usually the legs have less hair and the spider is usually a web spinner, although some are still hunters. One interesting aspect of spider legs is that they only have muscles for contracting the legs. To extend them, they use a hydraulic system where they push fluid into a leg to get it to move away from the body.

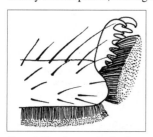

Figure 3: The claws at the tip of the tarsus are used by spiders to move through their webs.

Spider legs are usually covered with thick spines and/or fine hairs. These are all useful to keep the spider informed about its environment, either approaching prey or predators. Some very fine hairs detect wind movements and are very flimsy. For the advanced

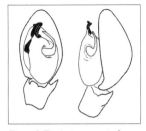

Figure 4: The last segment of a mature male spider's pedipalp is enlarged into a specialized reproductive organ. The same palp here is seen in ventral view (left) and lateral view (right).

Figure 5: The pedipalps are the male mating organ in the front of the body. On this spider, they are dark, whereas in others, most palps are similarly colored to the male's body.

arachnologists, the claws, pattern of spines, pattern of hairs, etc. are very useful for identification purposes. However, all these features require the use of a good quality microscope and much practice.

***Pedipalps***—The pedipalps are on the front of the spider's body in front of the first pair of legs and, typically, look like miniature legs. They are often called "feelers" by the general public, and their function is, indeed, mostly sensory as well as prey handling. Although the pedipalps look like legs, if you examine them closely you will see that they have six segments and not seven as do legs.

In most spiders, the pedipalps have no use whatsoever for identification purposes. As the spiders go through various molts, the pedipalp looks like a small leg. In the females, there are no drastic changes throughout her life. However, in the males, this structure develops into the reproductive organ (Figure 4). When the male enters the molt before maturity (called the "penultimate" molt), the tip of the pedipalp becomes swollen. At this point, no structures are present on the pedipalp, thus useful only to identify the spider is a male with one more molt to go before becoming fully mature. Once the penultimate male molts, the swollen bulb acquires various structures that are important in helping him align with the female's genital openings and

Figure 6: The male mating organ (palp) can be very simple or a very complex structure with many conspicuous features.

Figure 7: The great variation in male palps is used by arachnologists to designate and separate species. These palps (left to right) are from the hobo spider, giant house spider, and common house funnel weaver.

transfer sperm for mating (Figures 5 and 6, page 69). These structures are usually very consistent in form within a species but are very different among species. Because many spiders have poor vision, this is the mechanism that allows them to prevent mating with the wrong species in nature. These differences are also exploited by voyeuristic arachnologists who use the variation to separate species (Figure 7). Although it is most proper to call these palps, many people refer to them as "boxing gloves." Using the genitalia to separate species is beyond the scope of this book but the non-arachnologist can use them to identify a mature male.

*Chelicerae*—The chelicerae are the movable mouthparts in the front of the spider, the tip of which house the fangs. Just behind them are the mouthparts through which the spider intakes its nutrients.

The chelicerae are very useful for identifying the major groups of spiders. The mygalomorphs have parallel chelicerae such that they have to raise the front of their body upwards to swing the chelicerae outward, and unfurl the fangs in order to accomplish a downward strike into the prey (Figure 8a, page 71). The common spiders have opposing chelicerae that are more like pliers (Figure 8b, page 71). In order to envenomate prey, they spread their chelicerae sideways, unfurl the fangs, and bring them together like pincers. The hunting spiders typically have teeth bordering the furrow in which the fang rests (similar to a knife blade in a knife when it is sheathed). These teeth help the spider grab and hold on to its prey while the venom takes effect. Because web spinners typically use silk to wrap their prey, they usually lack these teeth.

**The Abdomen.** The abdomen is the second and more posterior main body portion of a spider. In many arachnids (e.g., scorpions, daddylonglegs), the abdomen is composed of segments, but almost all

 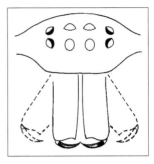

Figure 8a: The jaws of mygalomorph spiders attach to the front of the cephalothorax which requires them to lift their body and bite in a downward motion.

Figure 8b: The jaws of araneomorph spiders move in a side-to-side motion like a set of pliers which allows them to manipulate prey easily.

the spiders encountered by the pest professional will have unsegmented abdomens. However, abdominal segmentation occurs in very primitive, "missing link" spiders in Indochina, and one can find a few American mygalomorphs which have remnants of their segmented evolutionary past. Many spiders have stripes or chevrons running cross-wise over the dorsal surface of the abdomen; these are additional indications of the segmented past of their ancestors.

***Book lungs*** — The respiratory organs of spiders are called book lungs and are located on the underside of the abdomen toward the cephalothorax. They are usually lighter in color than the rest of the abdomen. They have slits where air enters, and the lungs contain sheets of tissue for oxygen exchange within the body. Because several sheets are piled on top of each other, they resemble the pages of a book and, hence, the name. This can be a diagnostic identifier because mygalomorphs have two pairs of book lungs (Figure 9), and common spiders have one pair (or in the smaller and more

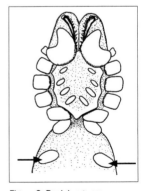

Figure 9: Book lungs are breathing organs consisting of saclike folds containing a series of sheet-like leaves.

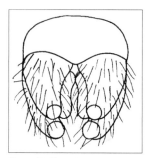

Figure 10: The spinnerets of a sac spider. Note they are conical in shape. Spiders exude silk through their spinnerets.

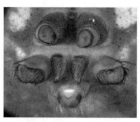

Figure 11: Most spiders have 3 pairs of spinnerets on the posterior portion of the abdomen. Silk is emitted from spigots on the surface of each spinneret. The unpaired structure at the bottom of the photo is the anal tubercle.

advanced common spiders, they have lost the book lungs and replaced them with a system of conduits called trachea, similar to insects).

***Spinnerets***—At the posteriormost portion of the spider's body, tube-like or conical structures can be found poking out (Figures 10 and 11). These are the spinnerets and are the organs through which silk is emitted. The ancestral spider had four pairs of spinnerets, but one pair of them has become vestigial or disappeared completely in the common spiders. Most mygalomorphs have lost yet another pair so they only have two pairs of spinnerets. Most spinnerets are small, but some spiders have spinnerets that can be about ⅓ the length of their bodies. The length or shape of the spinnerets is diagnostic for some spider families or genera.

The surface of each spinneret is covered with spigots which correspond to specific silk glands inside the body. The different silk glands have different purposes (Figure 12, page 73) and will be covered in the next paragraph. The placement of the spigots are specific to certain spinnerets; for example, in orb-weavers, the spigots that exude the main dragline silk are typically on the front spinnerets and the spigots that exude the sticky silk are on the hind spinnerets because as the spider moves through its web, it needs to be laying down the dragline silk first before it can cover it with sticky droplets for prey capture.

As mentioned above, most spiders have six spinnerets with one pair disappearing. However, in some genera, the fourth pair of spinnerets has evolved into a flat plate with many spigots. This plate is called a cribellum (Figures 13 and 14, page 73). Spiders that have a cribellum also have a row or two rows of special curved hairs on the metatarsus

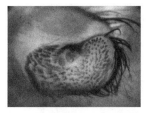

Figure 12: The surface of each spinneret is covered with spigots that emit silk. The different-sized spigots connect to separate glands that produce silk for specific purposes.

of the fourth leg, which is brushed over the cribellum to pull out the silk. These curved hairs are called the calamistrum. Creatures with these structures are called cribellate spiders. Their silk is not sticky like other spiders but instead look like puffy skeins of yarn that one would buy in a craft store. Although cribellate silk is not sticky, it acts like Velcro®, where the hooks and spines of insect legs get tangled up in the puffs of silk. This may be useful in identification because cribellate silk looks more like cotton candy and has a bluish tint to it. In comparison, the silk of non-cribellate spiders that use drag lines (like the black widow or the orbweavers) looks more like thin fishing line which glistens in the sun. One very frustrating aspect for the beginning arachnologist is that because the cribellum and calamistrum are used to spin silk, male cribellate spiders lose these structures upon maturity because, at that point, they abandon their webs to run around looking for females. In order to identify mature male cribellate spiders, one must first learn what the females look like and then extrapolate the other features of the spiders to the males.

***Silk Glands***—The evolution of silk in spiders is a topic that is exhaustively researched. Silk production is one of the evolutionary

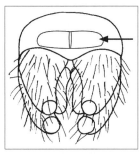

Figure 13: The cribellum is found above the spinnerets of cribellate spiders.

Figure 14: The cribellum is a flat field of spigots that evolved from a pair of spinnerets. It produces silk that ensnares by entanglement, not adhesion.

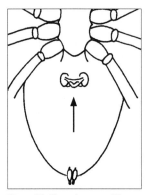 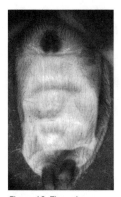

Figure 15: The epigynum of the female spider is on the underside of the abdomen (after Wegner).

Figure 16: The epigynum is the darkened structure near the top of the picture.

advances that have allowed spiders to become so successful in such a great diversity of habitats. Most spiders have four types of silk glands and each silk has a specific purpose. Some silk is used for making structural webs, another gland only produces attachment discs so the spider can attach other silk to the substrate. The egg sac is covered with its own special silk. Additional silk glands are found in the orbweavers and combfoot spiders where they produce sticky globules of silk that are laid down over the dragline silk in forming a web that filters the environment for flying prey.

*Reproductive System*—Just to the rear of the book lungs in the common spiders (or the anteriormost pair in the mygalomorphs) is a line that runs across the body. This is called the epigastric furrow. Just forward of this furrow between the book lungs is where the reproductive organs of spiders are found. In the male, these are of no diagnostic importance because once the male matures, he squeezes sperm out of an opening that is attached to the epigastric furrow and suctions up the sperm into his palp. However, this is where the female's reproductive organs lie and is the target for the male's palp during mating. In the more simplified spiders, very little indication of female genitalia occurs here; however, in the common spiders, the genitalia is usually a hardened plate called an epigynum (Figures 15 and 16). This structure consists of openings into which the male inserts structures of his palp and internal ductwork that carries sperm into the spermathecae where it is stored until the female extrudes eggs into an egg sac, fertilizing the eggs as they are passed out of the body. Just like the male palp, the female epigynum is typically very consistent in its appearance, and, therefore, is a diagnostic feature for arachnologists to certify a species' identity. However, this often requires the advanced skill of dissecting the area under a high quality microscope. For the pest professional, awareness of the epigynum may be just sufficiently useful to be able to identify a spider as a mature female. In some spiders, the epigynum is darkened and very identifiable; however, in other species, there may only be a small opening which may not be visible except under high magnification.

# CONTROL TECHNIQUES

*Kill a spider, bad luck yours will be*
*Until of flies you've swatted fifty-three.*
*(Author Unknown)*

Spiders are beneficial creatures and are responsible for keeping the numbers of many insect pests in check. Except for a few dangerous species, most spiders are relatively harmless and rarely come into contact with people. In fact, the reason that spiders are present in and around buildings is because insects that serve as their food source are present. The larger the population of insects that is present serving as a food source for spiders, the more spiders that are likely to be found.

Obviously, spiders such as the brown recluse, black widow, and brown widow require immediate control efforts when even one spider is found in a home or other building. The bites of these spiders are serious enough that customers may request extreme measures be taken to eliminate an infestation, if that is possible. Other common spiders found in homes, such as house spiders, cellar spiders, and jumping spiders, pretty much keep to themselves and require minimal control efforts—unless the building's occupant or owner specifically requests that all spiders be controlled in their home or business. Wolf spiders, by their very size and swift movements, cause such concern in many people that they want these spiders controlled even when only one is seen. Spiders which keep to the outdoors, such as the large garden spiders, require no control efforts because their presence in flower gardens is nonthreatening and many people see the beauty in their elaborate webs.

As with most structural pests, control effort and strategies for spiders depends much on the type of spider involved. Brown recluse spiders will require considerably more effort and different techniques to successfully control than will wolf spiders. Each situation will also be different due to the varying conditions, construction, and other factors unique to each building and its particular spider infestation. This chapter will introduce and describe in some detail the various techniques used to control spiders. These techniques will be mentioned when discussing the control measures for each type of spider included in this field guide. *Refer to this chapter for a detailed explanation of specific control techniques mentioned in other chapters of this field guide.*

**Identification.** In order to control any pest infestation, including spiders, you should identify the pest or pests involved, find where they live, and identify the conditions present that may be supporting the infestation. When dealing with spiders, the most difficult task will likely be correct identification of the spider involved. One of the primary purposes of this field guide is to assist pest professionals in the identification of the spiders found in and around structures. Until the spider has been identified, it may be difficult to determine all of the places where that spider might be harboring and identify any contributing conditions for that infestation. Until these items are known, successful control of the spider infestation may not occur.

It is important to instruct the customer to capture and keep a specimen of the spider(s) they see so that it can be identified. Collecting a specimen is most important in cases of brown recluse spiders. Many times, a homeowner is very concerned that this species is present. It must first be confirmed that brown recluse spiders are actually present before going through the extensive efforts commonly needed to control and, hopefully, eliminate them. Homeowners will misidentify almost any brown, medium-sized spider as a brown recluse, anywhere in North America including American states or Canadian provinces where brown recluse spiders are not known to exist.

Figure 1: Monitoring traps can be placed to capture spider species that move about in search of prey. This trap contains numerous brown recluse spiders.

*Monitoring Traps*—If the customer is unable to capture a specimen for identification purposes and is reporting that the spider was seen running on a floor or wall, *monitoring traps* may be placed in the areas where the spider was seen. These traps are only useful for capturing those spiders which actively hunt their prey (Figure 1)—spiders which build webs are typically found using a flashlight inspection. Leave monitoring traps out at least one night—two nights or more is better—and check them for specimens captured. This technique is heavily used in control programs for brown recluse spiders and will be discussed in the chapter dealing with that spider.

*Flashlight Inspections*—Simply taking the time to inspect using a flashlight is the best technique to find where spiders are living. As stated earlier, web-building spiders are easiest to find because the spiders are confined to those webs. Hunting spiders will be more difficult to locate due to the often secretive hideaways they use during the day. A *flushing agent* (pyrethrins aerosol) can be used to flush spiders, such as wolf spiders or the brown recluse, out of cracks and voids in which they might be hiding.

*Inspecting for Web-Building Spiders*—When dealing with web building spiders, check corners at the floor/wall juncture, corners of walls, and where the ceiling meets the walls. Also inspect behind and under furniture, corners inside closets, and behind toilets. Check the corners of windows both inside and outside as web-building spiders are frequently found around windows. Spiders thrive in windows because the sunlight entering through glass attract flying insects that have become trapped within a building.

Garages, crawlspaces, and basements are often the sites where the greatest numbers of spiders might be found because these areas are often undisturbed, are not cleaned regularly, and are likely to harbor larger numbers of insects which provide food for spiders. Spiders like the black widow often build webs among piles of items and boxes stored in these areas. Be sure to wear gloves when moving items to inspect for black widows. In certain cases, a damp, poorly ventilated attic will harbor many spiders because such attics are attractive to numerous insects. If the garage, crawlspace, basement, and attic are not inspected and included in the spider control program, success may be limited, depending on the situation. It is from these areas, as well as from outside, that spiders infiltrate to the living areas of the home or other building. Such key areas can provide a ready source of new spiders to replace the ones killed or removed inside the living areas.

Outside, inspect in corners of windows and doorways and under the eaves of the building. Also check behind gutters, under railings of decks, and where fences adjoin to the home. Any site where right angles are present can serve as a place for spiders to attach a web. A common place to find these spiders is around the light fixtures beside doors and other exterior lighting. Lights attract numerous flying insects which serve as the food for these spiders. Another possible site to find web-building spiders is among piles of debris or a firewood pile next to the structure. Black widow spiders are commonly found in such locations as well as behind thick vegetation against fences and walls.

*Inspecting for Hunting Spiders*—Active and passive hunters are more difficult to find because they do not have established webs where they spend most of their time as do the web-building spiders. Some hunting spiders, such as brown recluse, sac, and jumping spiders, spin silken web retreats in which to rest, however, these retreats can often be difficult to locate.

When dealing with jumping spiders, the usual place to find them inside is around window sills and doorways. Outside, jumping spiders will be found hunting for prey on the outside of the building, on decks, fences, and railings, and on shrubbery and tree trunks near the structure.

Wolf spiders are likely to be found resting in cracks under baseboards and in the bottom of closets inside. In garages and basements, this spider is likely to be found hiding among boxes and other piled items or in cracks around the garage door frame. Outside, wolf spiders prefer to rest under items such as stones, landscape timbers, firewood piles, and within leaf litter. They can commonly be found under decks and in crawlspaces resting or crawling on the walls or the floor. Wolf spiders (as well as other spiders) enter the living areas of buildings by following plumbing pipes, electric wires, and cables up into the structure.

Ground spiders are most common outside along the foundations of buildings and enter through exterior cracks, around windows, and under doorways. Inside, these spiders are most commonly found along baseboards, on curtains, and under furniture. Occasionally they can be found running or resting on walls.

Brown recluse spiders can be found anywhere in a building. Specific inspection tips will be described in the chapter discussing recluse spiders.

**Nonchemical Control Techniques.** For the control of most spiders in and around structures, nonchemical techniques will generally comprise the majority of the control efforts. To provide the best results, some or all of these techniques may be utilized in all spider control programs. Which techniques are used depends on the situation and the factors affecting that infestation. Insecticide applications are at best a short-term control measure, so if nonchemical techniques are not utilized, long-term results may not be realized.

Many of these nonchemical control measures may need to be completed by the customer. A written list of recommendations should be provided to the customer as well as an explanation why each recommendation is important.

*Spider Removal*—Since spiders are essentially beneficial creatures, capturing single specimens of certain types of spiders and releasing them outside is preferable to killing them. One jumping spider, crab spider, tarantula, funnel web spider, or other type of spider which seldom enters a structure can easily be captured and removed to the outdoors. Of course, spiders such as brown recluse spiders and black widow spiders should be killed.

Figure 2: Removing a single spider can be accomplished by using a cup and a piece of paper.

The easiest way to catch a spider is to place a jar or cup over the spider and slide a piece of paper under the top of the container so that the spider is trapped within it (Figure 2). Turn the container over while keeping the paper secure over the top, and take it outside and release the spider. A large spider like a tarantula can be coaxed into a box and removed.

*Sanitation*—Sanitation is likely the most important nonchemical technique when dealing with most structure-infesting spiders. Removal of actual or potential spider harborages and those harborages utilized by insects which serve as the spider's food source is the main sanitary concern. For example, if a building is experiencing an infestation of wolf spiders or ground spiders, removal of piles of debris, leaf litter, and heavy vegetation next the foundation will be very helpful in long term control of these spiders. If the building is a hospital, warehouse, or food plant, complete removal of all vegetation to create a 24 inch band next to the foundation is recommended. This band of bare ground should be either concreted or covered with plastic or a weed barrier with a 2-inch layer of small (e.g., pea) gravel placed on top.

Inside buildings, storing boxes off the floor and away from walls limits their use as harborage by spiders. Spiders, such as the brown recluse, however, may still crawl into and live in boxes stored in this manner. When dealing with this spider, each box may need to be opened and examined for spiders. Removal of as many cardboard boxes as

Figure 3: A cobweb duster is used to sweep spider webs from eaves, windows, and corners.

possible is helpful. If boxes cannot be moved, sealing all openings with tape after clearing the box of spiders can help keep them out. Be sure to wear gloves when inspecting in and around boxes to avoid accidental bites.

It is beneficial when dealing with web-building spiders to remove old webbing when and where possible. This serves two functions. First, removal of webbing allows one to easily determine new activity during future services. Second, and just as important, web removal gives the customer the impression of complete control. If a service professional simply killed the spiders and left the webbing, dust particles would soon collect on the webs, making them more visible to the customer. In their minds, customers may consider old cobwebs to be a sign of continued spider infestation. Any control program for web-building spiders should include web removal for these reasons.

Webs are most easily removed using a tool designed for that function. One such device is called a cobweb duster (Figure 3) and is available in most hardware stores. The device is equipped with a telescoping handle that allows the user to reach high ceilings and other high areas where webs might be found. (Figure 4). Cleaning of the device is accomplished by swirling it in a bucket of soapy water. If this device is not available, a broom can be used in its place although it is not as professional looking. During the inspection, it is best to carry a device to remove webs and also a nonresidual insecticide to directly treat spiders as these are

Figure 4: The telescoping handle allows the user to reach as far as 15 feet overhead.

found. Doing this saves time by avoiding the need to make two trips over the same area.

Clutter in storage areas, garages, sheds, and basements can provide numerous hiding or web-building sites for many spiders. The more clutter, the more spiders that are likely to be found because the clutter is also attractive to many insects that serve as food for spiders. Convincing the customer to straighten up this clutter and remove as much of it as possible can be helpful when dealing with many spiders. If brown recluse or black widow spiders are involved, straightening up and reduction of clutter may be critical to long term success. In these situations, the service professional may go through the stored items and treat any spiders found directly with a nonresidual aerosol insecticide before the customer straightens up. Boxes stored in these areas should be addressed as discussed earlier. Gloves should be worn to avoid accidental bites. Instruct customers to also wear gloves when moving or working with stored items.

Another consideration outside is to advise customers to cut the branches from trees and shrubs away from the roof and walls as spiders can crawl or jump off branches onto the structure.

*Exclusion*—As with most other pests which generally invade structures from the outside, preventing them from entering in the first place is the best course of action. Effective exclusion of all spiders, however, is very difficult due to the small size of the spiderlings that allows them to enter through the smallest of cracks. Larger spiders can be excluded by sealing cracks around win-

Figure 5: Xcluder™ fabric can be used to fill cracks and holes in order to exclude spiders and other pests.

dows and doors and other cracks in the exterior walls. Stuffing larger cracks with Xcluder™ fabric can exclude spiders, insects (Figure 5) and even mice. Doors and windows should be equipped with complete weatherstripping and soffit vents, foundation vents, and roof gable vents should have screens capable of excluding insects and spiders.

One of the authors conducted a study to determine the smallest size mesh screening that $2^{nd}$ instar black widow spiderlings could fit through.

He found standard screen mesh openings of 0.83 mm would allow spiderlings to crawl through, but the next size smaller mesh having 0.59 mm openings generally excluded these tiny instars. This study was completed after a large commercial retailer expressed desire to keep black widow spiderlings from entering through the screens attached to the large wall vents of a giant distribution warehouse. Screen consisting of 30-mesh size is recommended over 18- or 20-mesh to exclude spiderlings. Keep in mind, however, that $2^{nd}$ instar spiderlings of species other than black widows may be small enough to fit through even 30-mesh screens.

Some food processors and warehouses are located in areas next to or near fields, wooded lots, or rivers, lakes or other bodies of water. Such locations are havens for spiders producing untold thousands of spiderlings through the summer that may balloon onto the building. In these cases, it is important for these facilities to keep doors closed or use screening over doors and vents as described above.

*Exterior Lighting*—The presence of bright white metal halide lighting on the outside of some large commercial buildings can be a factor in some spider infestations. This type of lighting attracts large numbers of nighttime flying insects providing a ready food source for spiders along the walls of the building. Larger numbers of spiders outside invariably leads to numerous spiders entering the building. Changing the lighting to sodium vapor lamps that produce yellowish light attracts many fewer insects.

Homeowners can be advised to equip exterior light fixtures with yellow "bug light" bulbs.

*Ventilation*—Moisture reduction is a primary factor in controlling many structural pests, including spiders. Excessive moisture and high humidity conditions attract many insects which in turn attract spiders that prey on insects. Spiders, like most arthropods, are sensitive to moisture loss, therefore avoiding overly dry conditions. Installation of proper ventilation in crawlspaces and attics reduces moisture and consistently high humidity thus making the area less attractive to insects and spiders.

In crawlspaces, vents should be installed in as many exterior foundation walls as possible to provide maximum possible ventilation. When a crawlspace has low clearance or has excessive moisture, more foundation vents are necessary. It is also helpful to apply a plastic vapor barrier over the crawlspace floor. Follow state recommended guidelines and/or standard industry practices.

Attic ventilation is best accomplished by installing a sufficient number of soffit vents and *ridge vents* across the peak of the roof. This system works best because it provides air movement from all corners of the attic. Box vents, turbine vents, and/or attic fans combined with soffit vents may be less effective than ridge vents but are better than no vents at all.

*Vacuuming*—The use of vacuum devices to physically remove pests can be an important component in controlling many pests, including spiders. Vacuuming is especially useful in situations where large numbers of spiders and webbing is present because it takes less time to accomplish both of these tasks in these situations. Vacuuming will remove spiders, webbing, and egg sacs in one step. When only a few spiders are present, vacuuming may be more time consuming than using other techniques.

**Insecticide Applications.** Insecticides are rarely needed for many of the spiders that might be found inside buildings including jumping spiders, crab spiders, and light infestations of web-building spiders and hunting spiders. In many cases, nonchemical techniques alone will provide acceptable results. When dealing with the presence of larger numbers of spiders or dangerous spiders, like the brown recluse or black or brown widows, treatments can be a critical component of the overall control program.

Spiders are difficult to kill with insecticides primarily for the following reason. Web-building spiders generally spend a lot of time in their webs and little time in contact with treated surfaces. These spiders are not likely to pause on residual deposits long enough to absorb a lethal dose and die. Other types of spiders, such as wolf spiders, may allow their bodies to touch the surface below as they crawl and to remain for longer periods in contact with a treated surface. These types of spiders are more likely than web-builders to be controlled using surface treatments. Hairy spiders, like wolf spiders, are more likely to pick up residual dust particles and microcapsules due to the numbers of setae (hairs) on the legs and body. Too much reliance on treating surfaces where spiders crawl as the primary method of control, however, generally provides less than desired results. One of the age-old comments about spiders is that they don't get much of an insecticide load because they don't pick up enough pesticide because they don't clean their body parts. However, this is not true. Spiders routinely run their terminal leg and pedipalp segments through their mouthparts to clean them because the hairs are critical environmental sensors that alert the spider to its immediate environment. What may be more truthful is that the insecticides that work for insects have little effect on spiders. Looking through the literature, many studies show that there is great variation within one species as how different insecticides affect it (some being highly toxic, some having no effect). Similarly, regarding any particular insecticide, there is great variation among the species response where some species succumb immediately and other species are immune to the products.

Like most arthropod pests, spiders are best controlled by treating the areas where they live or rest. For web-building spiders, this means finding and treating the spiders in their webs. For hunting spiders, this strategy is a little more difficult to employ. The following describes many of the treatment techniques used for spiders and how they are applied. Refer to this section when these treatment techniques are mentioned in the control section of the chapters discussing specific spiders.

*Flushing*—The application of a nonresidual pyrethrins aerosol into likely harborages for spiders can flush out hiding spiders. This technique is primarily used when inspecting for spiders such as wolf or brown recluse spiders.

*Directed Contact Treatment*—This technique is commonly used on web-building spiders. Directed contact treatments involve applying a *nonresidual* insecticide directly onto an exposed spider (Figure 6). Insecticides such as aerosol formulations of pyrethrins are effective for such treatments. It is important to note that when applying these aerosol insecticides the application nozzle on the can be held at least 18-24 inches away from the spider to prevent the insecticide from liquifying on the wall behind the spider and causing stains or other damage.

Directed contact treatments are important because enough insecticide will definitely be applied to the spider to cause its death. As mentioned earlier, carry an aerosol nonresidual insecticide along during inspections so it will be handy when a spider is found.

Application of residual or nonresidual insecticides can also be made into cracks or voids to kill spiders harboring there.

Figure 6: Nonresidual aerosols can be applied directly onto spiders residing on or within their webs.

*Spot Treatments*—The application of a residual insecticide to small areas of surface can be effective if the treatment is properly placed. For example, the corners of windows, closets, and rooms where web-building spiders have frequently been found can be treated with residual insecticides. As these spiders spin their webs, they may remain in contact with the insecticide long enough to absorb a lethal dose. Applications to baseboard areas where hunting spiders are likely to crawl can be helpful in some situations involving these spiders. Such treatments, however, are not necessary in many spider infestations.

In unfinished basements, crawlspaces, and garages, the application of spot treatments using residual insecticides can be effective in *reducing* the numbers of spiders in these areas, but not necessarily eliminate them. SC, CS, or WP formulations generally work best for spot treatment applications. Apply the insecticide to sill plate and box header areas in unfinished basements and crawlspaces. Corners and floor/wall junctures in these areas can also be treated as web-building spiders will build webs there.

In garages, treat the floor/wall juncture and ceiling corners, and pay particular attention to the ceiling/wall juncture above the overhead garage door. If the garage does not have finished walls, the area between each wall stud will need to be inspected and treated as necessary because web-building spiders frequently locate their webs along such studs.

Outside, spot treatments can be used around windows and doors, around exterior light fixtures, under porches, and under eaves.

Avoid relying on spot treatments as a major part of a spider control program. When dealing with web-building spiders, sanitation and regular web removal may provide better long term results. For hunting spiders, sanitation, traps, crack & crevice treatments, void treatments, and directed contact treatments often provide the best results.

*Painting*—Some situations exist where liquid residual insecticides cannot be applied by spraying. In these situations, a concern usually exists that spraying will allow insecticide droplets to drift onto areas where insecticides should not be applied. In these cases, the insecticide can be applied using a paint brush or other type of brush.

The most common case where painting is necessary is in the awnings covering boats at a marina. Insecticides cannot be allowed to drift into the water surrounding the boats. A good way to treat such awnings is by dipping a cobweb duster (Figure 3, page 82) into a bucket of properly mixed, dilute insecticide, allowing the excess liquid to drip off and then spinning the device in the corners where spiders build webs. This method removes webbing and applies the insecticide at the same time.

*Crack and Crevice and Void Treatments*—When spiders are likely to be living in or enter through a crack or void, the best method for treating these areas is an application of a residual dust product (Figure 7). Dusts are ideal because they have a long residual life and a spider is likely to pick up more dust particles as it crawls through a crack than it would pick up particles of other insecticide formulations. As with insect pests, an application of a *light layer* of insecticide dust works best for controlling spiders. The best way to treat wall voids is to remove electric outlet covers and treat the void behind the outlet box. Be sure to use a plastic tip on the duster to avoid accidental electric shock.

Figure 7: Application of a residual dust can help in preventing spiders from entering inside.

A dust product containing silica gel or diatomaceous earth works well for spiders because these ingredients result in the spider dying from dehydration. Silica gel also remains effective for a year or more if applied into a dry area. Avoid using boric acid for spider control because it is a stomach poison and would require the spider ingest it to be killed, which is not likely to occur.

*Dusting of Crawlspaces*—The application of a residual dust insecticide to spider harborage areas in a crawlspace or attic can be more effective than spot treatments. Spot treatments, however, may work better in crawlspaces that are damp or wet.

When applying a dust in a crawlspace or attic, use a large hand-operated duster. Avoid using a power duster to treat the entire crawlspace or attic because too much dust will be applied than is necessary.

An inspection of the accessible areas of attics and crawlspaces needs to be made to determine where spiders are active. To save time, it is a good idea to take the duster along during the inspection and apply dust to active spider webs and spider harborages. Dusts work well when applied to webs because the spider will be in continual contact with treated surfaces. In crawlspaces, dust also may be applied to the sill plate areas and around piers. In attics, dust may need to be applied under or behind insulation when brown recluse spiders are found harboring there. Be sure to wear the appropriate safety gear when applying dusts in enclosed areas such as the crawlspace or attic.

*Space Treatments*—Space treatments may be the least effective control technique for spider control. Space treatments involve the application of nonresidual insecticides to the entire volume of space within a room or building. Since spiders are actually occupying a relatively small amount of this space—usually along wall and ceiling areas—most of the insecticide applied is wasted. Space treatments may be the only option in a few situations, however, such as a warehouse with a high ceiling or in certain crawlspaces. For a space treatment to be effective, the volume of the room to be treated should be calculated and the correct dosage of insecticide applied for that volume of space. Also, the air handling system and any pilot lights need to be turned off prior to the space treatment. Directed contact treatments, however, are generally more effective for spider control.

Figure 8: Treatments applied to foundations can be helpful when dealing with active hunters such as wolf spiders.

*Perimeter Treatments*—The application of a residual insecticide to the foundation and to ground vegetation next to a building can help somewhat limit the numbers of spiders which might enter a building (Figure 8). Perimeter treatments are most effective for hunting spiders which are generally located at ground level and travel along the wall/ground juncture. A perimeter treatment alone may not produce the desired results for many web-building spiders because spiderlings which are "ballooning" may fly over the treated areas directly onto buildings. Exclusion, treatment of exterior cracks that cannot be sealed, and spot treatments around windows and doors can help prevent these spiderlings from entering. A combination of control methods is likely needed for many situations.

*Fumigation*—Fumigation involves completely covering a building with tarps and introducing a penetrating gas to kill all of the insects and other pests inside the building. This technique is very expensive and is usually used only in severe infestations of brown recluse spiders. Fumigation requires specialized equipment and training and must be performed by a company licensed for fumigation work.

# IDENTIFYING SPIDERS

As with many pest control situations, the first step in dealing with an arachnid incident involves the correct identification of the spider(s). This field guide includes the families of spiders which a pest management professional is likely to encounter. These 24 families encompass those species that may be considered *urban* in that they come into contact with humans or their structures. The chapters that follow discuss, in detail, the key identifying characters, basic biology, key inspection tips, and key management strategies for each spider Family. Identification is limited to the Family level for two reasons: 1) Identification of spiders to species is often difficult and requires a good microscope and 2) Control measures are generally applicable to all species within the same family. Species of greater pest professional interest, such as the black widow spider, the brown widow spider, and the brown recluse spider, are discussed in their own separate chapters.

Illustrations of spiders appearing in each chapter include at least two drawings, most often a carapace drawing and a face drawing showing the eye pattern arrangement unless a different view is more diagnostic. Where appropriate, other line drawings are added which show a specific identifying character. Please note that whenever the size of a spider is mentioned, the measurement (i.e., ¼ inch) refers to the body only, from front to back, and does not include the legs because the body length is less variable and more easily measured than leg length.

When an arachnologist identifies a spider, it is usually dead and under a microscope so the physical features of the spider are used for identification. However, a pest professional typically will see the spider in its natural location and can glean much more information about the spider such as where the spider was found, whether it was in a web or not, what was the shape of the web or its egg sac or its molted skin. These are all very useful aspects that can be used for identification, sometimes allowing identification of the spider without the spider's presence.

This chapter provides a *taxonomic key* to help in identifying the Family of spiders to which a collected specimen might belong. The key deals mostly with the physical characters of a spider such as eye pattern arrangement, markings, and the shape of the spinnerets. The key also includes tips about biology and habits of the spider where these tips may be helpful in identifying a specimen.

**Using the Taxonomic Key.** This key was designed for the non-arachnological pest professional. Although a microscope would be very useful, a 16x or 30x hand lens should be more than sufficient.

The following is a brief taxonomic key to the spiders included in this field guide. To use this key, you must do the following:

1. Compare your specimen with the descriptions and illustrations in each couplet. Decide which of the two descriptions fits your spider.

2. If the *name of a spider* is located to the right of the dotted line in that couplet, that should be the identity of your specimen if it was identified correctly.

3. If a number appears at the right, proceed to the couplet following that number and repeat the process comparing your specimen with the descriptions and illustrations in the couplet.

4. Work your way through the couplets until your specimen is identified. Compare your spider with the color photographs in this field guide. Turn to the chapter which discusses your particular spider and compare it to the illustration and description included there.

If your specimen does not identify to any spider included in this field guide, you may need additional assistance. In this case, several specimens of the spiders should be collected, placed in a vial with 70% alcohol, and mailed to an entomologist for identification. Most states have a university with an entomology department that can be of assistance. However, be aware that it may require someone with actual arachnological knowledge to provide an accurate identification; trained arachnologists are sometimes hard to find. Since spiders are such a specially studied field within entomology, it would be a good idea to find out who in your state has experience with spiders and would be willing to identify specimens mailed to them. In some states, an amateur may actually have greater accuracy in spider identification because they have focused specifically on spiders whereas, university or county entomologists may only dabble in arachnology.

Finally, the banana spiders were left out of the key because you will probably never encounter one and if you do, it will be associated with bananas, a fruit handler, or a grocery store, so you can just go to that chapter immediately to confirm identification.

# TAXONOMIC KEY FOR COMMON FAMILIES OF URBAN SPIDERS

1. Chelicerae (jaws) are attached to the front of the cephalothorax (head area) and open or move directly up and down, parallel to the body, four book lungs on underside of the abdomen.
   *Suborder Mygalomorphae* –
   **MYGALOMORPH SPIDERS** → ........................ *(go to 2 below)*

1'. Chelicerae (jaws) are attached under the front of the cephalothorax (head area) and open or move from side to side, (in one species (woodlouse spider) the jaws are halfway between parallel and side-to-side), usually with two book lungs on underside of abdomen.
    *Suborder Araneomorphae* –
    **TRUE SPIDERS** → ......................................... *(go to 5 below)*

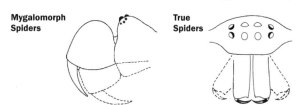

**Mygalomorph Spiders**

**True Spiders**

## MYGALOMORPH SPIDERS (2 THROUGH 4)

2. Large brown or black spiders with a body often up to or longer than two inches in length, very hairy, live in burrows in the ground, more common in southwestern United States.
   *Family Theraphosidae* – **TARANTULAS**

2'. Not large and hairy → ............................................. *(go to 3)*

3. Posterior (rear most) spinnerets are extremely long and extend out from the tip of the abdomen (see illustration), live in funnel shaped webs with sheet of silk web radiating out from funnel, found under rocks and in heavy vegetation.
*Family Dipluridae* – **FUNNEL WEB SPIDERS**

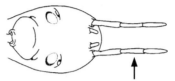

Long posterior spinnerets of the funnelweb tarantula.

3'. Spinneret much shorter and web is different than described above → ..................................................(*go to 4*)

4. Has very large endites (see illustration) present on the underside of the chelicerae (jaws), live in silk lined tunnel in ground which extends as a silk tube 8 to 10 inches above ground up a tree trunk or on top of the ground.
*Family Atypidae* – **PURSEWEB SPIDERS**

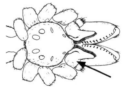

The endites under the chelicerae of the purseweb spider are very large.

4'. Endites are much smaller, not as described above, has rake-like area (see illustration) on side of chelicerae (jaws), live in silk-lined burrow in ground which has a single-piece lid or door at the top.
*Family Ctenizidae* – **TRAP-DOOR SPIDERS**

Chelicera of trapdoor spider showing rake-like rastellum.

## TRUE SPIDERS (5 THROUGH 23)

5. Has six eyes. ➜ ........................................................ *(go to 6)*

5'. Has eight eyes. ➜ ................................................... *(go to 9)*

6. Six eyes arranged in three pairs with a space separating each pair. ➜ ................................................................ *(go to 7)*

6'. Six eyes arranged in two pairs of three. ➜ ............. *(go to 8)*

7. Tan to brown spiders, may have violin-shaped marking on top of cephalothorax (head area), carapace flat, not humped.
*Family Sicariidae* – **RECLUSE SPIDERS**

7'. With variable coloration, carapace humped, not flat.
*Family Scytodidae* – **SPITTING SPIDERS**

8. Cephalothorax maroon, with large fangs in between parallel and side-to-side orientation.
*Family Dysderidae* - **WOODLOUSE SPIDERS**

8'. Light brown spiders that have six eyes arranged in two groups of three, very small spider, pale, long thin legs, short globe-like body, hangs upside down in webs in dark places.
*Family Pholcidae* – **SHORT BODIED CELLAR SPIDER**

(*Spermophora senoculata*)

**Carapace of brown recluse spider showing arrangement of six eyes.**

**Carapace of short bodied cellar spider showing arrangement of six eyes.**

PCT Field Guide **95**

9. Extremely long thin legs, oblong or round abdomen, eight eyes arranged in two groups of three with a pair of close-set eyes in between (see illustration), common in crawlspaces and basements where it hangs upside down in web.
*Family Pholcidae* – **LONG BODIED CELLAR SPIDER**

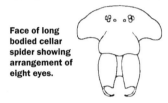

**Face of long bodied cellar spider showing arrangement of eight eyes.**

9'. Eye pattern not as described, legs not extremely long and thin, body and legs variable. ⟶ .....................................*(go to 10)*

10. At least two eyes much larger than the other eyes. ⟶ ..........
...............................................................................*(go to 11)*

10'. Eyes not as described above, all eyes more or less similar in size. ⟶................................................................*(go to 12)*

11. Middle two eyes on the bottom (anterior median) on the face are very large, small robust spiders with short thick legs, very hairy and often brightly colored, very quick and usually move by short jumps, active during the daytime.
*Family Salticidae* – **JUMPING SPIDERS**

**Face of jumping spider showing arrangement of eight eyes — middle eyes of bottom (anterior) row are very large.**

**Face of wolf spider showing arrangement of eight eyes — middle eyes of top (posterior) row are very large.**

11'. Middle two eyes, sometimes all four eyes of the top posterior row on face are large and round in shape, brown, sometimes hairy spiders, very fast runners which usually hunt at night (although some hunt during the day), female carries egg sac attached to tip of abdomen and spiderlings will ride on her back after emerging from the egg sac.
*Family Lycosidae* – **WOLF SPIDERS**

12. Eyes clumped in front portion of cephalothorax. ⟶ ............
................................................................................*(go to 13)*

12'. Eyes not clumped. ⟶ ...........................................*(go to 14)*

13. Eyes on raised turret, either solid tan with long thin legs or solid velvety brown or black.
*Family Filistatidae* – **CREVICE WEAVERS**

13'. Eyes flat on roundish cephalothorax, legs striped and held close to body if alive, small spiders rarely over 4 mm in body length.
*Family Oecobiidae* – **FLATMESH WEAVERS**

14. Six of the eight eyes in hexagon pattern with two smaller eyes underneath, abdomen pointed at tip, long thin spines sticking out perpendicularly from legs, active daytime hunters common on vegetation where they hunt and ambush their prey.
*Family Oxyopidae* – **LYNX SPIDERS**

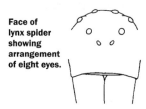

**Face of lynx spider showing arrangement of eight eyes.**

14'. Eyes not in a hexagon. ⟶ ......................................*(go to 15)*

15. Body crab-like, front two pair of legs longer than rear two pair, abdomen flattened, spider can move sideways like a crab, never in a web.
*Family Thomisidae* – **CRAB SPIDERS**

15'. Body not crab-like. → ..........................................(go to 16)

16. Combfoot on ventral surface of fourth tarsus (terminal leg segment), abdomen often globe-like.
*Family Theridiidae* – **COMBFOOT SPIDERS INCLUDING THE BLACK AND BROWN WIDOW**

Comb on last tarsal segment of the fourth leg of comb-footed spiders.

Face of comb-footed spider showing arrangement of eight eyes.

16'. No combfoot. → ..................................................(go to 17)

17. Chelicerae large and angling sideways toward tips, large fangs (see illustration), abdomen elongate.
*Family Tetragnathidae* – **LONG-JAWWED ORBWEAVERS**

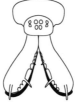

The long-jawwed orb-weavers have massive, curved chelicerae and fangs.

17'. Chelicerae not large and angled sideways. → ......(go to 18)

18. Abdomen angular (see Figure 8 on page 186), often with brushes of hair on front legs.
*Family Uloboridae* – **HACKLED ORBWEAVERS**

18'. Abdomen not angular. ➝ ......................................*(go to 19)*

19. Large pair of spinnerets consisting of one segment and tube-like looking like exhaust pipes sticking out from behind and often visible when viewed from above.
*Family Gnaphosidae* – **GROUND SPIDERS**

**Cylindrical shape of ground spider's spinnerets.**

19'. Spinnerets not large and are composed of two or more segments. ➝ ..........................................................*(go to 20)*

20. Posterior median eyes silverish and reflective, usually with many white hairs on the eye region of the cephalothorax, typically small spiders never larger than 5 mm in body length.
*Family Dictynidae* – **MESHWEB WEAVERS**

20'. Spider otherwise. ➝ .............................................*(go to 21)*

21. Abdomen large and usually bulbous or globe-like, sometimes with small humps, legs often covered with many spines, third pair of legs shorter than the others, often associated with an orb web.
*Family Araneidae* – **ORB-WEAVERS**

21'. Abdomen not large and globe-like. ➝ *(go to 22)*

22. Large spiders with eye pattern as shown in illustration, resembles wolf spider but less hairy, female carries egg sac under body by grasping with her mouthparts.
*Family Pisauridae* – **NURSERY WEB AND FISHING SPIDERS**

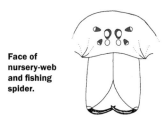

Face of nursery-web and fishing spider.

22'. Spider not as above ⟶ *(go to 23)*

23. Posterior spinnerets longer than other spinnerets, live in sheet-like web, eyes usually in a 2-4-2 pattern (see illustration) or for *Tegenaria* spiders eyes in two slightly curving rows (see illustration), coloration usually a combination of tans and browns with striped or leaf-like patterns on abdomen.
*Family Agelenidae* – **FUNNEL WEB WEAVERS**

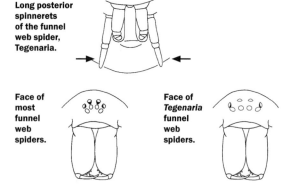

Long posterior spinnerets of the funnel web spider, Tegenaria.

Face of most funnel web spiders.

Face of *Tegenaria* funnel web spiders.

23'. Posterior spinnerets short and conical, eyes usually small and widely spaced from each other, typically abdomen and cephalothorax are only one color although the colors can be different between the two body parts.
*Families Clubionidae, Corinnidae, Miturgidae –*
**SAC OR RUNNING SPIDERS**

**Conical shape of sac spider's spinnerets.**

# MYGALOMORPH SPIDERS

Figure 1: The fangs of a mygalomorph rest side by side so the spider has to strike outward like a rattlesnake.

Figure 2: The venom glands of a mygalomorph are housed inside the chelicerae. In this photo the sidewall of the hardened chelicera is dissected away so the white venom sac is visible.

The spiders belonging to the infraorder Mygalomorphae include the tarantulas and their relatives. "Mygale" comes from the Greek word meaning "mouse," which was in no doubt given to the large tarantulas because of the similarities of their hairs to that of the fur coat mice. The morphological aspects that designate a spider as a mygalomorph include: 1) parallel fangs (also referred to as "orthognath") which resemble that of a rattlesnake, being side-by-side and requiring the spider to strike outward at its prey (Figure 1); 2) two pairs of book lungs, visible on the underside of the abdomen; 3) four spinnerets, whereas most common spiders have six (although some mygalomorphs do have six spinnerets); and 4) venom glands housed at the base of the chelicerae instead of in the carapace (Figure 2).

These spiders are robust with thick legs, often with strong spines, bends, or angles in the front legs of the male, which he uses to hold off the female during courtship. The eight eyes are very small and clustered together in the front of the carapace. Some, like the tarantulas, are extremely hairy whereas others are glossy and smooth. Although mygalomorphs are typically large spiders, the smallest ones are $1/10^{th}$ of an inch at maturity.

Mygalomorphs have slow-metabolisms and are long-lived. Many inhabit burrows for most of their lives and can secure them with a lid, a fold, and, in the more elaborate, an actual trapdoor which the spider holds in place with its fangs (trapdoor spiders). Others make a silk "purse" that runs up and down the trunk of tree where the spider feeds by attacking prey that

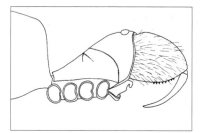

Figure 3: The chelicerae of mygalomorph spiders attach at the front of the cephalothorax.

lands on the outside of the purse, whereupon it is pulled inside after the spider makes a slit in the silk, which is later repaired (purseweb spiders).

Because of the orientation of their fangs (Figure 3), in order for them to catch prey, mygalomorphs must raise up the front part of their bodies to get enough clearance for a downward swing.

These spiders are not typically a concern to the pest professional; however, because of their large size and robust legs, homeowners typically are anxious by their presence and request control measures. In one encounter with a particularly pugnacious species in Northern California, a homeowner actually called the police to have the spider removed. Most mygalomorphs encountered by your clients are wandering males looking for mates, but because the males will soon perish, little can be done to control them. However, in heavily populated mygalomorph areas, multiple males can be seen running through a yard during mating season, and, for the homeowner, it does get somewhat disturbing when they have to fish several specimens out of an in-ground swimming pool with the unexpected event that some of the waterlogged spiders may recover and start walking again.

# TARANTULAS

*Family Theraphosidae*

**Key Identifying Characters.** As you can imagine, it is fairly easy to identify a tarantula. Not many spiders look like these creatures and no other ground-dwelling spider grows near to it in size. Tarantulas are mygalomorphs and, as such, the arrangement of their fangs is in parallel configuration similar to that of a rattlesnake where they extend their fangs forward to strike at prey and then secure it by pulling the fangs back toward its body. Tarantulas are primitive spiders so they have two pairs of book lungs, as opposed to the more evolutionarily specialized spiders, which have only one pair. These will be evident on the underside of the abdomen. However, parallel fangs and four book lungs is a characteristic of all mygalomorphs (which would also include purse-web and trapdoor spiders). The diagnostic feature that further differentiates tarantulas from other mygalomorphs is that tarantulas have a heavy mat of dense hairs called scopulae on the ventral surface of their tarsi which helps them move and hold on to prey. Most North American tarantulas are black or brown and typically covered with hairs over the entire body except for the posterior of the abdomen if the spider has rubbed off its urticating hairs (see below). The most common species found are of the genus *Aphonopelma*. When mature, tarantulas range from 9 mm to about 75 mm in body length with two genera and 55 species in North America.

Tarantula.

**Biology in General.** Tarantulas, similar to all mygalomorphs, are primitive spiders, therefore, they have slow metabolisms. Whereas many common spiders go through their life cycle in a year or less, tarantulas are very long-lived. After they hatch in the maternal burrow, the spiderlings stay with their mother and may even live there for a year before dispersing on their own. The spiderlings opportunistically take up residence in available burrows as they do not have cheliceral musculature strong enough to excavate a burrow of their own. Eventually, they get big enough to modify a burrow in which they stay for several years. Around buildings, a tarantula may even take up residence in a suitable space between and under rocks or other items.

Males mature in five to seven years after which they leave their burrows in search of females. These wandering males are the spiders that are often seen running across roads, sometimes in great numbers. If you happen to find such a male, it would be best to leave him alone. He will not make a good pet because he only has a few weeks to live. It would be better to just let him continue and wish him luck in finding a female so he can make several hundred more babies like himself. Females can live up to 20 years, even longer in captivity. Common spiders may molt every few weeks, but tarantulas are typically restricted to one molt per year. To do this, they lie on their backs to extricate themselves from the old exoskeleton (see *Care of Spiders*, Figure 5, page 61). It is a slow process that takes several hours.

Most of the North American tarantulas live in the southwestern quadrant of the United States, typically in dry habitats like desert or chaparral. They make their burrows in the ground, enlarging them as they grow. The burrow doesn't have a well-defined cover as seen with the trapdoor of trapdoor spiders but, instead, the spiders cover their burrow entrance with a thinly-woven patch of silk which is typically covered with detritus.

The wandering male finds a female's burrow and entices her to vacate the burrow so the mating ritual can occur. After mating, the male may walk off in search of more females, which is not unlikely as oftentimes burrows of females can be relatively dense in supportive habitats. The female lays an egg sac about 2 to 3 inches long inside the burrow containing 400 to 800 eggs.

**Biology in Regard to Pest Control.** Wandering males looking for mates will be the main interaction between a tarantula and a homeowner.

**Toxicity.** North American tarantulas do not have very toxic venom; their bite is said to be similar to a bee sting. Considering the docile nature of American tarantulas and that even tarantulas freshly collected in the field can be handled without incident, if you are ever bitten by a North American tarantula, you more than deserved it.

However, tarantulas are not without risk. As covered in the chapter on *Health Aspects of Spiders*, tarantulas have minute, barbed hairs on the end of their abdomens which they can displace toward a threatening stimulus. The hairs can lodge in the nasal cavities and eyes and cause severe irritation.

**Pet Trade Tarantulas.** Although a pest professional would rarely have the chance to interact with an exotic tarantula (Figure 2), in the event a tarantula escapes its cage and crawls off the property of the owner, as one could imagine in areas where tarantulas are not a common sight, a highly-excitable neighbor may request eradication if they are unlucky enough to score a visit from an escaped beast.

There is a greater health risk from pet trade tarantulas, which are typically exotic species originally from South America or Africa, which are reared in captivity and legally sold in North America. Their venom is more toxic and because they usually live in trees rather than in underground burrows, they have urticating hairs that are more aerodynamic (Type III hairs), and seem to be more willing to use this defense.

Figure 2: Many people keep exotic tarantulas as pets, and it is conceivable a pest professional might encounter one that has escaped its cage.

**Inspection Tips.** Tarantulas seldom enter the living areas of buildings although they may enter warehouses, factories, or even garage areas when doors are left open at night or cracks are present of sufficient size. Most of these intrusions are during the late summer or fall when the males are actively searching for females. It is possible for tarantulas to burrow next to a building's foundation or even in a crawlspace. These spiders, however, most likely will never enter the building. If a tarantula was seen wandering inside and then disappeared, it will most likely be found hiding behind or under some piece of furniture, box, or other item. In garages, check behind or under stacks of wood or other items.

Even an experienced tarantula hunter will have difficulty finding the burrows of tarantulas. Greater luck will be had when finding the occasional tarantula burrowing under a rock, a log, or a piece of debris. One of the authors discovered a tarantula in Texas living inside the partially buried skull of a dog located in a field.

**Key Management Strategies.** Generally, unless it's a situation where males are moving in large numbers, calls to control tarantulas are limited to single individuals that have entered a structure or are found on the patio, entranceway, or near the foundation. Capture and removal of the spider to an area away from the building is the best course of action. Killing a spider that is gentle in nature and an important part of the surrounding environment is not warranted.

A tarantula can easily be captured by coaxing it to crawl into a box. The spider can then be easily transported to a site away from the building where it was found.

*Sanitation*—Removal of debris, wood piles, etc. away from the foundation walls of the structure or from the property eliminates potential burrowing sites for tarantulas. This can reduce, but not eliminate, the possibility that a tarantula will be seen outside by the homeowner.

*Exclusion*—In cases where the threat of tarantulas entering is a common occurrence, sealing cracks and crevices at the foundation level of the building may be necessary. Ensure tight fit of foundation vents and weatherstripping on doors. Stress the need to keep doors closed tightly in areas where tarantulas have been a problem.

Tarantulas will live in abandoned rodent burrows so make certain that all burrows are located and filled in.

Insecticide applications should rarely, if ever, be needed to control tarantulas.

# OTHER MYGALOMORPHS

**Key Identifying Characters.** Several additional families of spiders occur in the mygalomorphs other than the theraphosid tarantulas. These include the purseweb spiders (Atypidae), trapdoor spiders (Ctenizidae), and funnel-weaving (Mecicobothriidae). Most of these spiders are large but some can be only 1/10th" in body length. They are typically robust spiders with thick legs but are less hairy than tarantulas; some have cephalothoraces that are hairless and quite shiny. There may be several physical adaptations like a raised cephalothorax to accommodate digging muscles of the mouthparts and rasp-like addendages to aid in scraping dirt to excavate burrows. However, most of these spiders are fairly similar looking. It would require a microscope to distinguish the necessary differences among the species, but it would not be worth a pest professional's efforts to do so because these spiders are not of great pest control concern. These other mygalomorphs can be lumped into one large category.

**Biology in General.** Mygalomorphs spiders exhibit quite an interesting diversity of lifestyles and prey capture although mostly they are passive ambush predators. The purseweb spider builds a silken tube up the trunk of a tree and covers it with some detritus. The spider then takes up a position inside the tube. An insect may land on the outside of the tube, assuming it is a tree trunk, the spider moves in the tube under the prey, impales its fangs through the silk to stab the prey, makes a hole and pulls its lunch through the hole, after which it then repairs with silk (Figure 1). The spider then dines inside.

Many burrow dwellers will take up residence inside a pre-existing burrow and then line it with silk and expand the burrow as it grows. The entrance to the burrow may be a raised turret, a flexible collar which is closed during the day and opened at night when the spider is waiting for prey, may be covered with a simple layer of detritus, or may be a specialized hinged trap-

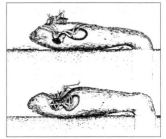

Figure 1: Purseweb spiders bite through its purse-like web to capture its prey.

Figure 2: Trapdoor spider.

door which is camouflaged to match its surroundings. In the case of the trapdoor spiders (Figure 2), they sit in the burrow and hold the trapdoor with their fangs as one can see two fang holes on the inside of the door if it is forced open (Figure 3, page 111).

The genus *Cyclocosmia* has a vertically-flattened posterior with hardened grooves. If chased by a predator into its burrow, it will run to the bottom of the ever-constricting burrow until it gets to the point where its abdomen wedges against the sides of the burrow. The predator then runs into the hardened flat end of the spider and is thwarted in its attempt to turn the spider into dinner.

In almost all of the cases, during the night, the mygalomorph spider waits in the entrance of the burrow. Trip lines may be present radiating away from the burrow to help alert the spider to presence of a potential meal. If vibrations from the ground or the silk triggers a response, the spider rushes out with lightning speed, grabs the passer-by with its fangs, and takes it into the burrow where it is consumed.

**Biology in Regard to Pest Control.** Most of the interactions with the pest professional will be males that are out searching for females. In a few reported cases, large numbers of wandering males of various species have been found in pools or pool filters. However, such cases are rare.

**Toxicity.** These large spiders have massive fangs and strong cheliceral muscles so if they were to bite someone, it would be significantly painful from the physical penetration of the skin alone. In one incident in Georgia, a cyrtaucheniid spider (genus *Myrmekiaphila*) penetrated a garden glove and punctured a finger during a bite. Although the highly toxic Sydney funnel-web spider, *Atrax robustus*, from Australia is found in this group, none of the North American species have toxic venom for humans. Even for their prey, they often subdue their conquest with power cheliceral muscles rather than venom potency and the prey may attempt to crawl away while it is being consumed by the spider.

**Inspection Tips.** Spiders such as the purseweb and trapdoor spiders are rarely found around structures, being more common in wooded areas. Encounters by homeowners will typically involve a single male as it wanders in search of a mate. Time spent looking for burrows or webs is not necessary.

**Key Management Strategies.** Usually, the pest professional will be asked to identify a single spider captured by the customer. As this is often the only such spider involved, no treatments are necessary.

In the event that larger numbers of males are wandering into pools or garages, little can be done in the way of treatments to deter such wanderings. Typically, such larger migrations result around homes located in more rural locations.

Figure 3: A trapdoor spider's burrow. The trap door usually closes under its own weight.

# TRUE SPIDERS

## *Infraorder Araneomorphae*

Most of the spiders found in the world belong to the infraorder Araneomorphae. They were given a common category name of "true spiders" which is a rather unfortunate moniker because that means the other infraorder (the mygalomorphs which include the tarantulas) are excluded. Occasionally, someone will say they heard that tarantulas aren't "true spiders." Well, yes and no. They are spiders but are not found in the araneomorph group which are known as the "true spiders." However, a much better term for araneomorphs would have been the "common spiders" as they are incredibly numerous in populations, in the number of species, and in just about every spider category.

Araneomorphs have several characteristics that bind them as a group. First, their fangs come together side-to-side like the grabbing end of a pair of pliers (Figure 1); they are opposing rather than like the forward-directing, rattlesnake-type fangs of a tarantula. (However, in some species like the woodlouse spider, the fangs are halfway between parallel and opposing [and are called paraxial], but they are still considered in the araneomorphs.) Araneomorphs have, at most, one pair of book lungs or some species are small enough that they have merely an internal set of tubing called tracheae which are used for respiration. Araneomorphs also have their venom glands housed in the cephalothorax and have long ducts that deliver the venom from the glands to the fangs at the tip of the chelicerae. Another trait of these common spiders is that they have a quick development time; they can go from spiderling to a massive, large female in about six months, growing quickly, molting every few weeks, and becoming mature.

Tremendous diversity is seen with the araneomorph spiders in all aspects of their being. Some have eyes that rival humans for their visual acuity; others use their eyes merely to determine whether it is day or night. Most are nocturnal,

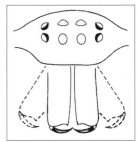

Figure 1: The chelicerae of the true spiders attaches under the cephalothorax and move side-to-side.

some are diurnal. They build webs to catch prey, or they hunt it down and pounce upon free-living creatures. Some have long, thin, fragile legs, others are stocky and built like little tanks. They live under rocks, under bark, on top of water, in vegetation, in burrows, in your house, and in your garage. One genus in Europe lives underwater, and jumping spiders have been collected at 20,000 feet on mountains in Asia. They crawl, jump, run, and the smaller ones will be carried by uplifting air currents where they may travel miles before landing and have been collected floating around at 15,000 feet in the air. There are not many places on earth, except the frozen poles, where araneomorph spiders are not successful at survival.

# IMPORTED OR BANANA SPIDERS

## *Families Ctenidae, Sparassidae*

**Key Identifying Characters.** Spiders considered in this chapter are not of major concern to the pest professional. Occasionally, however, a large spider from a cargo shipment may get loose in a grocery store or at port while cargo is being unloaded which may instigate a frantic request for pest control. Almost all of these spiders are large, fast moving creatures and almost all have come in on banana shipments. Because of the chance of major misidentification of a harmless spider as a potentially deadly one, information is provided here.

Figure 1: Giant huntsman spider, *Heteropoda venatoria*.

The most common spider brought into this country is a large huntsman spider, *Heteropoda venatoria* (Sparassidae) (Figure 1). The females are mostly brown and the male is more gaudily colored with tans and white (Figure 2, page 116). The key identifier is that each sex has a white "moustache" between its chelicerae and its eyes. Its legs are rotated backward by 90° and the spider moves somewhat like a crab. This spider is pantropical in distribution. It is found in Florida, Hawaii, Central America, South Africa, Taiwan, and Hong Kong, but most of the specimens get to this country in bananas from Ecuador. In Hawaii, they are referred to as "cane spiders" and are household pests. This giant huntsman spider ranges from 22 mm to about 28 mm in body length when mature with a common eye pattern of two rows of four eyes (Figure 3, page 116).

Another group of spiders occasionally brought into this country are of the genus *Cupiennius* (Ctenidae) (Figure 4, page 117). These are long-legged brown spiders with a body length of about an inch and a leg span of 3 or more inches. The spider of greatest concern is brown and has bright red hairs on its chelicerae and is known as *Cupiennius chiapanensis* (no common name) (see photo in color section of book).

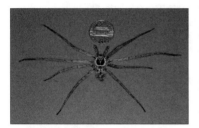

Figure 2: *Heteropoda* spiders are large as can be seen by the comparison of this male spider with a U.S. penny. Females are even larger.

The importance of this spider is not that it is dangerous but that it is frequently misidentified even by experienced arachnologists as a potentially deadly Brazilian spider (see below). Another species (*C. getazi*) often is brilliantly covered with orange hairs and the ventral surface of each femur is white with black polka dots. There are a few other species, but they are only rarely transported. These spiders are basically restricted to Central America and most of the specimens get to this country in bananas from Costa Rica and Guatemala. The *Cupiennius* spiders range from 15 mm to about 25 mm in body length when mature.

**Biology in General.** These spiders are free-living hunters that live in the leaves of tropical plants. Because bananas are washed thoroughly before shipping it is thought that the spiders get into the cargo during the packing process in the warehouse, not in the picking of bananas.

**Biology in Regard to Pest Control.** There isn't much for the pest professional to know about these spiders. The important aspect is that the spider with the red cheliceral hairs (*Cupiennius chiapanensis*) is all too often mistaken for the potentially dangerous armed spider (*Phoneutria* sp.) from Brazil. Because sufficient literature is not available to North American arachnologists or, more likely, because arachnologists may only be asked to identify these spiders once in a career, the necessary literature is not readily available. One can imagine a frantic dock foreman standing in the doorway of an arachnologist's office desperately awaiting an answer as to whether or not this large spider is dangerous. All too often, the initial verdict that a

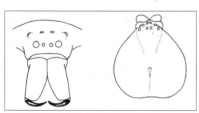

Figure 3: Face and carapace of the huntsman spider.

potentially deadly spider may have been found sets off a whole cascade of anxiety-filled actions such as the refusal to unload thousands of dollars of perishable cargo which then may decay and lead to great financial loss, having to protect a crew of dock workers, developing measures to prevent further importation of these spiders, and convincing employees that the risks to their survival are low or non-existent.

Much of this identification can be determined by geography and commercial practices. The potentially dangerous armed spiders live in Brazil. The most feared (*Phoneutria fera*) lives in the central Amazon, far away from commercial banana plantations and export activity. Brazil does grow bananas, but their crop is consumed domestically so few bananas are exported. Most of the bananas brought to the U.S. come from Central America where armed (*Phoneutria*) spiders are very rare or from Ecuador. From either of these places, the spider would be *P. boliviensis,* the smallest *Phoneutria* species which is not known to be very dangerous. Therefore, the chance of encountering a highly dangerous armed spider in the U.S. is virtually zero.

**Toxicity.** Huntsman and *Cupiennius* spiders have bitten people, but despite the large size, their venom has virtually no effect on humans. Bite symptoms typically last an hour or two.

**Inspection Tips.** Encounters with banana spiders usually involve a single spider that has arrived in a box of bananas or has wandered indoors (*Heteropoda venatoria* in Florida or Hawaii). These spiders can be lightning fast as they try to get away from you. The spider will quickly run to cracks or clutter or behind and under furniture to hide. Inspection involves using a flashlight and moving items in an attempt to locate the spider.

In the event that the spider cannot be located, large monitoring traps or glue traps may be placed along walls behind furniture, boxes, or other items.

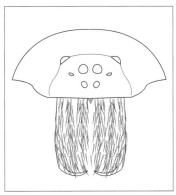

Figure 4: Face and carapace of *Cupiennius*.

**Key Management Strategies.** These spiders generally require no treatment applications as single spiders are typically involved. Single spiders inside can be captured by clapping a bowl or large cup over the spider and sliding a piece of stiff paper or cardboard (see Control Techniques, page 77) underneath to trap the spider.

Where banana shipments are involved, the spider should be killed and kept in 70% alcohol as a specimen for training purposes, if so desired. However, because they are not dangerous, they also can be collected alive and easily maintained in a large aquarium or plastic gallon jar with a screw top lid if provided with water and prey such as crickets, although you might want to make sure that the spider cannot escape.

In Florida or Hawaii where *Heteropoda venatoria* may occur outside, the spider may be released outdoors away from the building. If the spiders are common around a particular building, steps such as the following may be taken to minimize spiders occurring indoors:

- Pruning tree and shrubs away from the walls and roof of the building.
- Sealing exterior cracks and openings.
- Ensuring tight-fitting weather-strips on the bottom of doors (including the garage).
- Installing screening on all foundation and attic vents.
- Eliminating piles of items around the structure or moving these to the property's perimeter.

# CREVICE WEAVER SPIDERS

## *Family Filistatidae*

**Key Identifying Characters.** There are several ways to identify crevice weaver spiders, which for the pest professional is one of the few species of the genus *Kukulcania*. From a gross morphology standpoint, the female has a large body with a robust abdomen and a velvety brown or jet black coloration (Figure 1). They sometimes are mistaken for small tarantulas. The males are long legged and light tan (Figure 2); this is one of the most common spiders misidentified as a brown recluse spider. The male also has extremely long palps (the mating organs) in the front part of his body, much longer than most other spiders. Most people who see these spiders would never suspect that the tan male would be the mate of the dark female. On closer inspection, crevice weavers have eight eyes all clustered on a small turret (Figure 3, page 120), looking very similar to that of a mygalomorph spider (Figure 4, page 120).

Crevice weaver spider.

Another characteristic identification feature is the web. The typical web of a crevice weaver around a home is started in a hole or small cavity. If it is in a hole in a flat wall, the spider will make its web out of

Figure 1: The female crevice weaver spider is large bodied, velvety in texture, and dark brown or black in color.

Figure 2: The male crevice weaver is thin-bodied, long-legged, tan, and looks nothing like the female. This spider is frequently submitted as a brown recluse.

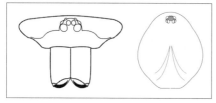

Figure 3: Face and carapace of crevice weaver spider.

fluffy silk, with crooked lines of silk extending out in a large circle around the hole (Figure 5, page 121). The most common species is *K. hibernalis* which lives throughout the southeastern quadrant of the United States; in the west, there are several species, most notably *K. utahana* and *K. arizonica*. The genus is restricted to the southern third of the United States. Crevice weaver spiders range from 5 mm to about 18 mm in body length when mature with three genera and seven species in North America.

**Biology in General.** These are nocturnal spiders, which wait in the entrance of the web at night for prey to get entangled in its web whereupon the jiggling of the silk alerts the spider to presence of dinner. The silk is dry and entanglement is more like Velcro® adherence; the catching strands have no adhesive. These spiders can go monumental periods without eating. One vial containing a crevice weaver was misplaced in the laboratory for nine months, yet when rediscovered the spider was thin but still alive.

**Biology in Regard to Pest Control.** Crevice weavers can become very common in and around some homes and outbuildings with many individuals being found within a short distance from one another.

**Toxicity.** There have been a few recorded bites from *Kukulcania* spiders with minor symptom development.

**Inspection Tips.** A single crevice weaver spider may wander inside, but homeowners typically notice them outside on the walls of the home, shed,

Figure 4: The eyes of male and female crevice weavers are all clumped together toward the front of the body and look similar to that of tarantulas.

or other building. Usually, it is the webs that are most noticeable.

If this spider is found indoors, check for the webs associated with a hole or crack in exterior walls (Figure 6). Also, turn over items in contact with the ground and inspect under piles of items such as firewood or lumber near the home or building.

Figure 5: Crevice weavers make a characteristic web usually emanating from a hole in a structure.

**Key Management Strategies.** Individual spiders can be removed when found inside. Ideally, it is best to capture the spider and release it outside. This can be done by placing a cup over the spider and sliding a paper underneath. Release the spider on the ground in a landscaped area away from the building.

To help prevent future invasions by crevice weavers, keep heavy vegetation cut away from the foundation, and seal cracks and holes in the building exterior in which this spider might construct webs.

The only treatment that might be needed is an application of a residual dust into cracks where the spiders might be hiding followed by having the customer seal these openings afterward.

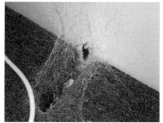
Figure 6: Although not often found indoors, crevice weavers will usually be found in a crack or hole with their characteristic web surrounding it, such as seen with this one found in a furniture store.

# JUMPING SPIDERS

## *Family Salticidae*

**Key Identifying Characters.** If there were ever a diplomat which could best change people's opinions about spiders, it would be the jumping spiders. They are agile, little, short-legged, squat, tank-like creatures that are colorful, animated, and cat-like in their hunting behavior. One could go so far as to calling them cute. Many people who don't like spiders do like jumping spiders. They come in a rainbow of colors, mostly the males who use their coloration as part of species recognition during courtship behavior. They may be festooned in brilliant cherry red or bright greens, yellows, stripes, spots, etc.

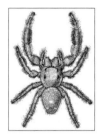

Jumping spider.

The diagnostic characteristic of the jumping spiders is that on the front vertical surface of their face, they have only the four eyes of the anterior eye row with the middle pair of eyes looking like large headlights (Figure 1). In comparison, most of the other families of spiders use their eyes predominately to let them distinguish day from night or to clue them into the approaching shadow of a possible predator. Jumping spider eyesight, however, is acute and used actively in hunting prey.

Because they are diurnal, jumping spiders are often noticed by homeowners. The biggest jumping spiders are in the genus *Phidippus* which also has some of the more colorful species. The bold jumping spider (*P. audax*) can grow up to 18 mm long, is jet black in color with conspicuous white dots on its abdomen. When nearly mature, the juveniles sport orange or yellowish dots instead. Several *Phidippus* jumping spiders have brilliant red coloration on their abdomens and, sometimes, both cephalothorax

Figure 1: The eyes of a jumping spider are diagnostic for the family. It is the only spider family with large anterior median eyes.

and abdomen. These spiders look like velvet ants (wingless wasps that pack a wallop of a sting) and, therefore, gain protection via mimicry. Many of the *Phidippus* jumping spiders have iridescent green chelicerae where the color does not come from pigmentation but rather specialized reflective structures that selectively reflect green wavelengths of light (see color image section of the book). The zebra jumping spider (*Salticus scenicus*) is common throughout the eastern United States with its very distinctive, striped metallic coloration.

Figure 2: The jumping spider has excellent vision.

Jumping spiders are the most diverse family of spiders worldwide with over 5,200 species which is 12% of the world's known species of spiders. In North America, jumping spiders range from 1 mm to 22 mm in body length when mature with 63 genera and more than 315 species.

**Biology in General.** With their excellent vision in the anterior median eyes (Figure 2), jumping spiders are strictly diurnal in their activity cycles. These large eyes are an amazing adaptation in that their visual acuity approaches human eyesight. If one were to hold up a hand with fingers outstretched in front of a restrained spider, the image that would be recorded from the spider's retina would show five individual fingers. Also, although the front lens of the jumping spider is fixed, they have muscles in the back of their eyes that allow the spider to focus in on prey. This vision is so acute that one Pacific species which preys on other jumping spiders is able to determine whether its potential prey is looking at it or looking away; the predatory jumping spider freezes when the stalked jumping spider peers toward it and advances when it turns.

Jumping spiders use their vision to stalk and attack prey. Because they can see from a good distance, they carefully approach their prey in short, intermittent steps. When they get close enough, they make a final cat-like pounce and grab prey with their fangs. If the spider jumps through the air to snag prey on a distant branch, it will attach a strand of silk so that if it misses the prey or the branch, it will swing in the air until it can pull itself back up to safety. They can also prey on non-moving items such as arthropod eggs and plant nectar.

Another aspect of their superb vision and their brilliant coloration is that they use both of these in their mating display. The usually colorful male will face the female and start a display which can include a variety of bewitching romantic overtures, such as leg and palp waving or circling, flicking of its tarsi, flaring of its chelicerae, drumming with his palps and front legs, all of which, of course, the female may find irresistible. Because their vision is so keen, this sensual semaphore code allows the female to determine whether or not this male is the correct species, which prevents matings between two incompatible individuals.

Although they do not use silk for prey capture, jumping spiders do make a silk retreat under bark or in rolled up leaves. This is also where they lay their egg sac, which the mother guards. Some spiders actually spend the night hanging by a strand of silk from a tree. By doing this, they are safe from nocturnal predators including other spiders that might be patrolling for a meal.

**Biology in Regard to Pest Control.** Jumping spiders are free-roaming hunters so they rarely venture inside buildings, rather spending their daytimes running around vegetation and the outside of houses.

**Toxicity.** The larger jumping spiders have painful bites, but the pain is mostly mild venom along with strong cheliceral musculature that imbed fangs into one's skin. The large *Phidippus* jumping spiders are most likely to inflict bites. Bites from jumping spiders are mild and self-limiting.

**Inspection Tips.** Jumping spiders will readily enter buildings from the outside. Often, this invasion is a result of following prey through cracks, under doors, through windows, etc. Once they have gained entry, jumping spiders quite often become permanent guests. These spiders will most often be found hunting around window sills and doorways because of the numbers of insects that are likely to be attracted to windows. Silken retreats may be found behind curtains, under furniture, between books on shelves, etc. although the retreats are often difficult to locate.

Outdoors, jumping spiders live in brightly-lit areas around the perimeter of buildings. Inspection of the walls of the building, around windows and doors, on decks, on fences and railings, and on shrubbery and tree trunks near the structure will often reveal a few jumping spiders. One technique that can be used to quickly determine if a spider is a jumping spider is to put your finger down about a foot in front of it and begin to draw a circle around it. If the spider turns to follow the movement, it most likely is a jumping spider.

**Key Management Strategies.** Most of the efforts to control jumping spiders center around controlling insect pests that they may be feeding upon. The role of the pest control professional is twofold: 1) identify the food source and gain the customer's cooperation in ridding the structure of these other pests, and 2) make necessary recommendations for excluding future infestations.

*Spider Removal*—Because they are highly beneficial creatures, single jumping spiders found inside should be preserved by capturing them and releasing them outside. Place a cup over the spider and slide a piece of paper under the cup to trap the spider inside. The spider can now be carried outside and released.

*Exclusion*—Though often impossible to totally exclude all spiders and insects that invade structures, any amount of exclusion efforts are helpful in preventing infestations. Seal cracks around windows and doors and other cracks in the exterior walls and ensure doors and windows have complete weatherstripping. Soffit vents, foundation vents, and roof gable vents should have tight-fitting screens of small mesh size.

*Sanitation*—Any sanitation efforts that will reduce insect populations will likely help reduce spider populations. Removing any potential harborages for insects such as debris, leaf litter, firewood piles, and heavy vegetation will be beneficial. Cutting the branches of trees and shrubs away from the roof and walls helps keep insects off the building that jumping spiders might hunt as food.

*Insecticide Applications*—Insecticide applications are seldom needed to control jumping spiders. By far, the best method for keeping jumping spiders out is the sealing of potential cracks where they could enter. Insecticides can be used in limited instances, however, as described below.

*Flushing.* The application of non-residual insecticides into cracks or crevices around windows and doors may help to chase out jumping spiders and insects that may be hiding in them.

*Directed Contact Treatment.* Direct application of a nonresidual insecticide to an exposed jumping spider will kill that one spider.

Removal of the spider as described previously is better than treatment.

*Spot Treatments.* The application of a residual liquid insecticide (WP, SC, or CS) to small surface areas may be effective in minimizing jumping spider invasions inside. Areas which might need treatment include around doorways and windowsills and around other entry points that cannot be sealed, such as weep holes on brick buildings.

# COLOR IDENTIFICATION GUIDE

Tarantula, *Aphonopelma* sp.

Wafer Trapdoor Spider, *Aptostichus* sp.

Banana Spider, *Cupiennius chiapanensis*

Banana Spider, *Cupiennius getazi*

Armed Spider, *Phoneutria boliviensis*

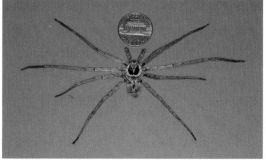

Huntsman Spider, *Heteropoda venatoria*, male

Huntsman Spider, *Heteropoda venatoria*, face

Crevice Weaver, *Kukulcania* sp., female

Crevice Weaver, *Kukulcania* sp., male

Crevice Weaver, *Kukulcania* sp., female, black form

Jumping Spider, *Habronattus* sp., female

Jumping Spider, *Habronattus pyrrithrix*, male

Bold Jumping Spider, *Phidippus audax*

Jumping Spider, *Phidippus otiosus*, female

Jumping Spider, *Phidippus otiosus*, female

Johnson Jumping Spider, *Phidippus johnsoni*

Johnson Jumping Spider, *Phidippus johnsoni*, male

Zebra Jumping Spider, *Salticus scenicus*

Jumping Spider, *Menemerus semilimbatus*

Wolf Spider, *Schizocoza mccooki*

Wolf Spider, *Rabidosa* sp.

Wolf Spider, *Hogna helluo*

Wolf Spider, *Hogna helluo*, carrying egg sac

Wolf Spider, *Hogna helluo*, with babies on her back

Fishing Spider, *Dolomedes triton*

Fishing Spider, *Dolomedes* sp., female

Spitting Spider, *Scytodes* sp., male

Spitting Spider, *Scytodes* sp., female

Parson Spider, *Herpyllus propinquus*

Ground Spider, *Cesonia trivittata*

Ground Spider, *Urozelotes rusticus*

Mouse Spider, *Scotophaeus blackwalli*

Sac Spider, *Trachelas* sp.

Woodlouse Spider, *Dysdera crocata*

Yellow Sac Spider, *Cheiracanthium mildei*, female

Sac Spider, *Trachelas pacificus*

Green Lynx Spider, *Peucetia viridans*

Green Lynx Spider, *Peucetia viridans*

Crab Spider, *Xysticus* sp.

Yellow Crab Spider

Brown Recluse, *Loxosceles reclusa*, female

Brown Recluse, *Loxosceles reclusa*, female (L), male (R)

Brown Recluse, *Loxosceles reclusa*, mother with offspring

Brown Recluse, *Loxosceles reclusa*, note the faintness of the violin marking

Mediterranean Recluse, *Loxosceles rufescens*

Garden Spider, *Argiope aurantia*

Golden Silk Orb-weaver, *Nephila clavipes*

Basilica Orb-weavers, *Mecynogea lemniscata*, mating

Shamrock Spider, *Araneus trifolium*

Marbled Orb-weaver Spider, *Araneus marmoreus*

Orb-weaver Spider, *Neoscona oaxacensis*

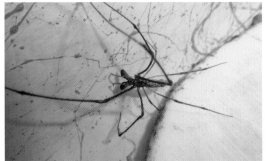

Silver Long-jawed Orb-weaver, *Tetragnatha laboriosa*

Spiny Orb-weaver, *Gasteracantha cancriformes*

Cellar Spider, *Pholcus* sp., eye arrangement

Cellar Spider, *Holocnemus* sp., female with egg sac

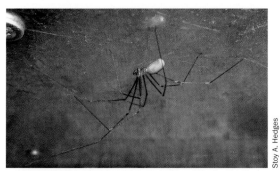
Cellar Spider, *Pholcus* sp.

Funnel Weaver, *Hololena nedra*

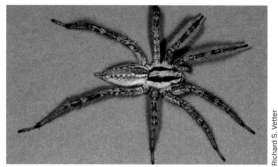

Funnel Weaver, *Agelenopsis aperta*, female

Hobo Spider, *Tegenaria agrestis*

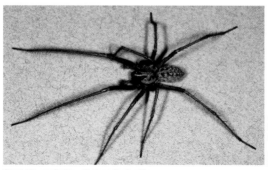

Giant House Spider, *Tegenaria duellica*

House Spider, *Parasteatoda tepidariorum*

Comb-foot Spider, *Steatoda triangulosa*, female

False Black Widow, *Steatoda grossa*, female

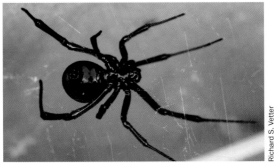
Southern Black Widow, *Latrodectus mactans*, female

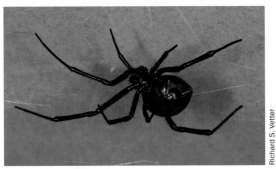
Western Black Widow, *Latrodectus hesperus*, female

Brown Widow, *Latrodectus geometricus*

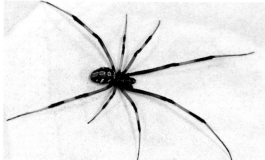

Southern Black Widow, *Latrodectus mactans*, male

Western Black Widow, *Latrodectus hesperus*, male

Brown Widow, *Latrodectus geometricus*, male

Southern Black Widow, *Latrodectus mactans*, immature with dorsal spots

Flatmesh Weaver, *Dictyna calcarata*, male

Meshweb Spider, *Oecobius navus*

# WOLF SPIDERS
## *Family Lycosidae*

**Key Identifying Characters.** The most diagnostic feature that allows one to identify a wolf spider is the unique size and arrangement of the eyes. Whereas most spiders have eight eyes in somewhat of two rows of four, wolf spiders have what is referred to as a 4-2-2 pattern (Figure 1). A straight or almost-straight row of four small, similarly-sized eyes are located just above the margin of the carapace, above that, two very large "headlight" eyes face forward, and then to complete the arrangement, a second row of two large eyes sit back about ¼ of the distance of the carapace where they are directed up and outward (Figure 2, page 130). Wolf spiders range from 3 mm to about 35 mm in body length when mature, with 16 genera and 238 species in North America.

Wolf spider.

Beyond this diagnostic feature, wolf spiders are typically covered in various shades of tan, brown, and black. The coloration can be strikingly discrete as if stripes were applied with a fine paintbrush, or it can be diffuse with no pattern. Colored rings, spots, or faint stripes may be present around the legs. They can range from very small spiders about 3 mm in body length, up to the gigantic *Hogna carolinensis* which can be an inch and a half long in body, and up to 3 inches from leg tip to leg tip. They have three claws on their tarsi (two large, one very small)

Figure 1: The wolf spider has excellent vision with large posterior median eyes.

Figure 2: Eye arrangement on a wolf spider.

which is a characteristic of web spinning spiders so this is probably a vestigial feature from when they maintained webs for prey capture. As an indicator of such, one genus of wolf spiders, *Sosippus*, makes a web like the funnel-weaving spiders, but they are not a concern to the pest control industry.

**Biology in General.** Wolf spiders are active hunting spiders; they move around, lying in wait for prey to cross their path, and then jump out and subdue it, thus the origin for their common name. They use their excellent vision to track their environment, which is why they can see you coming and run to avoid you. They can be both nocturnally and diurnally active. Wolf spiders have a reflective organ in their eyes, which allows them to see at night. An amusing activity in wolf spider areas is to go out with a flashlight at night, get close to the ground, holding a flashlight low and horizontal. If the beam of light hits the wolf spider's eyes, you will be able to see the eye-shines reflected back at you; for the larger species, this may be visible from up to 30 feet or more.

Another semi-unique feature of wolf spiders is that the female carries her egg sac attached to her spinnerets (Figure 3). This increases the risk for predation on the female because she is slower moving with her sac and also, the sac is a brilliant white color, which makes the cryptic brown spider easier to track visually when she is running across leaf litter or the forest floor. Once the babies are about to hatch from the egg sac, the mother needs to open the sac for them because they cannot escape on their own.

Figure 3: The wolf spider female carries around her egg sac attached to her spinnerets. She has to open the egg sac for the babies so they can escape. Then they crawl on her back and ride with her for a week or so.

Once this happens, the babies climb upon the mother's back and ride on the top of her abdomen for a week or two before they fall off and start life on their own (see the color image section of the book). If the spiderlings are knocked off, they quickly reassemble on mom's back before she runs off.

Wolf spider habitat can vary from extremely wet areas to desert; however, the majority of wolf spiders seem to require a water source to survive. If a wolf spider is placed in a vial, it needs a wet piece of paper towel or cotton wick or else it will die within a day or so. Although most are active wandering hunters, one genus (*Geolycosa*) lives in a burrow, enlarging it as it grows and captures prey that ambles past its opening.

Wolf spiders have very acute vision; therefore, their mating behaviors are quite interesting to watch as the male does his mating dance of amorous arm-waving semaphore and quick vibration of his abdomen which causes a buzzing (called "stridulation"). Taxonomists typically use the reproductive organs of spiders to differentiate species. Because wolf spider mating is directed by visual means, little evolutionary pressure has occurred for their reproductive organs to differentiate in order to maintain species recognition; therefore, a lot of similarities exist among the hundreds of species. The end result is that wolf spider taxonomy is still in a state of flux and names are still being changed. One of the biggest changes is that the genus *Lycosa* used to apply for many species in North America, but now that genus is restricted to the Eastern Hemisphere. Most of the North American "*Lycosa*" species are currently being transferred to other genera, particularly a new one named *Hogna*.

**Biology in Regard to Pest Control.** Wolf spiders are solitary hunters and enter homes as incidental pests unless populations are so high that they routinely saunter through a home. Homeowners have, at times, been surprised when after spraying females carrying babies, the babies scatter. The homeowner usually interprets this as the spider "exploded" and babies came pouring out of the inside of the body. Wolf spiders will almost always be found on the ground, rarely on walls and ceilings.

**Toxicity.** Wolf spiders will readily bite if trapped; however, most of the effects are likely going to be due to the mechanical pain of fang puncture and their strong cheliceral musculature, not venom toxicity. Nonetheless, because some of these creatures can attain a large size, the pain may be significant but short-lived.

**Inspection Tips.** Wolf spiders are normally outdoor spiders although, as indicated, they will enter buildings by following or hunting prey. A single spider seen indoors crawled in from outside so inspection needs to be done both outdoors and indoors. That single spider may be the only wolf spider that has actually entered the home, however, so the inside inspection may take longer with less to show for it than an outdoor inspection.

*Monitoring Traps*—Monitoring traps are very useful in determining the extent of a wolf spider infestation. These sticky traps are also more likely to capture and remove the offending spider(s) than will an indoor treatment. These traps are also important to use if the customer is unable to capture a specimen, particularly in those areas where brown recluse spiders are prevalent. A customer's description of a spider will not give you the exact identity, so a spider needs to be found or captured in order to identify it and select the correct control measures.

Traps should be placed in the vicinity of the sighting, near or beside doors, in crawlspaces, near garage doors, patio entrances, foundations, and window wells outside. Placement along walls under furniture and other items usually provides the best results. The traps should be checked after one or two nights.

*Flashlight Inspections*—Using a flashlight and a flushing agent (pyrethrins aerosol) may help to locate wolf spiders that might be resting in cracks and crevices, under baseboards, and in the bottom of indoor closets. In garages and basements, they may be found around boxes and other stored items. Outside, they may be found under stones, landscape timbers, firewood piles, under decks, and under leaf litter. In crawlspaces, wolf spiders will be found under items on the crawlspace floor, in cracks between wooden timbers, or simply running along walls, sill plates, or the floor.

**Key Management Strategies.** Because wolf spiders are active hunters that are most commonly found outdoors, extensive treatment efforts indoors are rarely necessary. The best long-term strategy is to focus on sanitation and exclusion to both reduce insect populations on which spiders may feed and to prevent them from entering.

*Sanitation*—A key ingredient in a program for wolf spiders is the reduction of insect populations by removing their harborages. Removal of debris and leaf litter from around the perimeter of the structure together with cutting vegetation away from the foundation and walls of the building are very helpful towards this goal. Where possible, elimination of thick ground covers, such as ivy or lirope (monkey) grass, near the foundation is beneficial in minimizing overall numbers of spiders and insects near the building.

*Exclusion*—Wolf spiders are most likely to enter under doors or through cracks around windows. They will also enter from the crawlspace by following wires and pipes. Sealing as many cracks as possible in exterior walls and around plumbing leading from the substructure is beneficial. Doors and windows should be equipped with complete weatherstripping, and soffit vents, foundation vents, and roof gable vents should be equipped with tight-fitting screens.

*Spider Removal*—Because wolf spiders are beneficial creatures, preserving them by capturing a single spider found indoors and then releasing it outside is a noble goal. A cup can be placed over the spider and a piece of paper slid underneath to entrap the spider. The cup can then be taken outside to release the spider.

*Exterior Lighting*—Metal halide lights on the outside of structures attract insects which, in turn, attract spiders, including wolf spiders. Changing to sodium vapor lights on commercial buildings attracts fewer insects, thereby reducing spider populations. Installing yellow "bug" light bulbs in the outdoor fixtures of homes will also attract fewer nocturnal insects.

*Insecticide Applications*—Nonchemical control strategies provide the best long-term results for controlling this spider. In situations where numerous wolf spiders are present or where they are seen repeatedly inside, insecticide applications may be required to provide the desired results for the customer. The amount and type of treatments needed will depend on the situation.

*Directed Contact Treatment*—If a wolf spider is found and cannot be removed by trapping it or by vacuuming, it can be killed by applying a short burst of a nonresidual aerosol insecticide directly onto it.

*Crack & Crevice and Void Treatment*—Indoor wolf spider retreats are likely to be located in cracks at or near the baseboard level. When an infestation involving many spiders is encountered a light application of a residual dust insecticide into cracks may be beneficial. Cracks and holes in the exterior should also be treated with dust insecticide, including weep holes. Treatment of these cracks should especially be done before they are sealed.

*Spot Treatments*—When wolf spiders are known to be in an area but their retreats cannot be located, spot treatments using a residual liquid insecticide can be helpful. Application of a WP, SC, or CS formulation works best. Spot treatments should be made to floor/wall junctures in rooms where spiders have been seen or trapped. These treatments are best made to areas under or behind furniture or other items because spiders are more likely to spend more time in protected sites like these.

Monitoring traps should also be utilized in cases where spot treatments are necessary. Remember, each spider caught on a trap is one fewer the customer might see.

*Perimeter Treatments*—When numerous wolf spiders are found indoors or when repeated invasions of wolf spiders occur, a perimeter treatment with a WP, SC, or CS formulation may be helpful. This treatment, by itself, is not likely to provide good results unless other steps such as sanitation and exclusion are completed. The application should be made to the foundation and up to several feet away from the building. Landscaped areas with heavy ground cover vegetation should also receive close attention.

# NURSERYWEB AND FISHING SPIDERS

## *Family Pisauridae*

**Key Identifying Characters.** Spiders of the family Pisauridae are known as nurseryweb spiders because of the maternal care the females exhibit for their egg sacs (genus *Pisaurina*) or fishing spiders because they live near bodies of fresh water and occasionally prey on fish (genus *Dolomedes*). They look similar to wolf spiders in coloration, with tans and browns predominating, but their eyes are more equally sized (Figure 1). These spiders can get rather large with some species being able to span one's palm.

Nurseryweb spider.

The nurseryweb spider (*Pisaurina mira*) is the most common pisaurid found in homes in the eastern United States. It has a broad stripe on the cephalothorax and abdomen. Another related species (*P. brevipes*) is similar-looking and is occasionally found by homeowners, prompting calls to pest management professionals. Among fishing spiders, a huge species (*D. tenebrosus*) sometimes enters homes. Females can have a body up to 26 mm in body length; this spider occurs from Florida north to Maine. Another species (*D. triton*) has vivid white lines and conspicuous white spots on its abdomen. Pisaurid spiders range from 5 mm to 37 mm in body length when mature, with three genera and 14 species in North America.

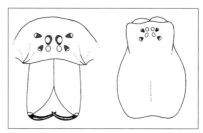

Figure 1: The fishing and nurseryweb spider look like wolf spiders but their eyes are more equal in size.

Figure 2: The nurseryweb spider female carries her egg sac under her body.

**Biology in General.** Nurseryweb spiders got their name because the female carries her egg sac around underneath her body (Figure 2). Sometimes, she has to walk on tiptoe to move around with the bulky batch of babies. When the spiderlings are getting ready to emerge, the mother will fold over the leaves of a plant—poison ivy is a favorite—and place the egg sac within the protective leaves. She will protect her egg sac until the spiderlings hatch and disperse.

Fishing spiders are typically found near a source of slow or non-moving water, such as streams and ponds. They sit at the edge of the water with the front legs dangling using the ripples to detect insects and, sometimes, fish that break the surface (Figure 3). They can also run across the water, using the hairs on their legs to stay above the surface. Nurseryweb spiders are found further from water and can be collected in the woods.

**Biology in Regard to Pest Control.** Nurseryweb and fishing spiders are solitary hunters and rarely enter homes although they could readily be found around waterfront property, boathouses, or recreational outlets where boating and fishing occur. People find it somewhat disturbing to have a massive spider scoot out from underneath a canoe when it is overturned for use.

**Toxicity.** Despite their large size, very few bites from this spider are reported and tissue damage from bites has been minimal.

**Inspection Tips.** Nurseryweb spiders prefer the outdoors and will only occasionally enter

Figure 3: The fishing spider can walk on the surface of water and hunts fish and aquatic insects.

a building while following or hunting prey. Their large size and coloration makes them easily confused with wolf spiders. Unlike wolf spiders, however, infestations of nurseryweb spiders will usually be confined to one or two specimens indoors. Capturing a specimen for identification is important to the selection of control measures.

If the customer is unable to capture a specimen, monitoring traps placed in the vicinity where the spider was seen may be the most efficient inspection method. Inspecting with a flashlight behind and under items located at floor level may also be useful. Monitoring traps should be checked after one or two nights.

Outside, inspect under items lying on the ground such as landscape timbers, rocks, logs, and piles of debris, and under decks, doghouses, and sheds.

**Key Management Strategies.** Because the nurseryweb spider is found so infrequently inside buildings, the best approach to follow is removal of potential harborages near the building and then sealing potential entry points.

*Spider Removal*—Because nurseryweb spiders are beneficial creatures, preserving them by capturing a single spider found indoors and then releasing it outside is a preferred disposal method. A cup can be placed over the spider and a piece of paper slid underneath to entrap the spider. The cup can then be taken outside to release the spider.

*Sanitation*—The important thing in controlling nurseryweb spiders is to eliminate as many potential harborages as possible that could be used by the spiders and their insect prey. Debris, leaf litter, heavy vegetation, and piles of items should be removed or cut back from the foundation.

*Exclusion*—Sealing as many exterior cracks around windows and doors and in the exterior walls as possible is a key to preventing inside invasions of this spider. Doors and windows should be equipped with tight-fitting weatherstripping, and soffit vents, foundation vents, and roof gable vents should be equipped with tight-fitting screens.

*Insecticide Applications*—Normally, nonchemical control measures are sufficient for controlling nurseryweb spiders and interior applications of insecticides should rarely be necessary. In fact, it is likely that you will be called upon to identify a single specimen that has been caught by someone and that will turn out to be the only spider that has entered. If numerous spiders have managed to enter a building, treatments outside may be required.

*Directed Contact Treatment*—A nonresidual aerosol insecticide can be directly applied to the nurseryweb spider in order to kill it although this is generally not advised or necessary. Removal of the spider by capturing it as described above and releasing it is a better approach.

*Crack & Crevice and Void Treatment*—Treatment of cracks and holes in the exterior walls using a residual dust insecticide before sealing them is a good strategy for preventing nuseryweb spider invasions.

*Perimeter Treatment*—The application of a residual WP, SC, or CS formulation to the foundation and the ground next to a building should only be necessary if numerous spiders have managed to enter the building. With nurseryweb spiders, this situation should rarely be encountered.

# SPITTING SPIDERS

## *Family Scytodidae*

**Key Identifying Characters.** Spitting spiders are easily identifiable because of their eye pattern, which is similar to the recluse spiders, a close relative. They have six eyes with a pair in front and a pair laterally on either side with a gap between the pairs (Figure 1). Because of this eye arrangement, they are often mistaken for recluse spiders. However, spitting spider coloration is distinctly different from that of recluse spiders (Figure 2, page 140). Spitting spiders have spots all over the body including the legs or can be completely black. Also, in order to accommodate their spitting glands (see below under Biology in General), the cephalothorax is greatly enlarged and humped toward the rear (Figure 3, page 140). No other spider looks like this. They also have thin spindly legs. The most common species is *Scytodes thoracica*, which is widespread. Spitting spiders range from 3.5 mm to 10 mm in body length when mature with one genus and seven species in North America.

Spitting spider.

**Biology in General.** Spitting spiders are wandering hunters. They are unique among North American spiders in that they have enlarged venom glands that have been modified for spitting a sticky substance which tacks down its prey so it can't escape. When a spitting spider encounters prey, it will aim its fangs and rotate them outward in a quick alternating fashion, spitting out silk

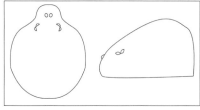
Figure 1: The spitting spider has an eye pattern similar to brown recluses (six eyes in pairs).

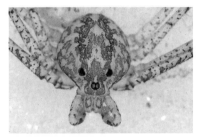

Figure 2: Although the eye patterns of spitting and recluse spiders are similar, the body coloration of a spitting spider is very different.

Figure 3: The specialized venom glands spit out a sticky secretion. When seen in side view, the cephalothorax is humped to accommodate the oversized glands.

and laying down a zigzag line similar to a sewing machine stitching thread. The prey is fastened to the substrate and the spitting spider then can approach safely to bite and feed. Although the cheliceral silk glands were modified from venom glands, the sticky secretion itself is not toxic. If you were to remove the spider after it tacks down the prey, the prey do not die. The female spitting spider carries the egg sac underneath her body.

**Role of Biology in Control.** These spiders are solitary hunters, often synanthropic, and exist throughout most of the continental United States.

**Toxicity.** A published study regarding nine spitting spider bites shows that they are harmless despite their close taxonomic relationship to recluse spiders.

**Inspection Tips.** Spitting spiders occur infrequently inside homes but when found inside, often a breeding population can become established such that spiders are seen on a somewhat regular basis.

Pest control inspections for these spiders is not easy as they often lie motionless for extended periods tucked into a crevice, behind furniture, or along molding (Figure 4, page 141). Often, they remain above ground level, being seen more frequently on the wall or ceiling than the floor.

Monitoring traps placed at floor level may catch a spider or two but usually inspection with a flashlight is a better option.

**Key Management Strategies.** Treatments indoors are typically not necessary for spitting spiders. Usually a homeowner will catch a specimen to identify, and this is the only spider found. In those cases where the customer sees more than one spitting spider or sees them on a more regular basis, then further control efforts may be warranted.

One of the authors experienced a spitting spider infestation that remained in his own home over several years. Spitting spiders would regularly be seen, as many as one to two dozen per year. Usually the spiders would be noticed when resting or crawling across lighter-colored walls. Such spiders would be captured for a collection or be released outdoors. One year, the gutters needed to be replaced on the home along with over 100 feet of water-damaged fascia board. In some areas of the soffits' interiors, numerous silverfish and an occasional spitting spider was observed. Ultimately, after the fascia and gutters were fully repaired and the area dried, interior sightings of both silverfish and spitting spiders ceased and became rare events.

This case may be indicative of some spitting spider and even recluse infestations inside homes. A chronic or underlying moisture problem permits the survival of a key spider food source, like silverfish, that permits a long-term, chronic spider infestation.

Figure 4: Spitting spiders often sit motionless along moldings and corners and can be difficult to spot.

Although great numbers of spiders are not seen within living spaces, they appear regularly enough to become a concern to the occupants. Determining whether such conditions exist and correcting them may prove to be the solution to a chronic infestation.

*Sanitation*—Removal of debris and leaf litter from around the perimeter of the structure together with cutting vegetation away from the foundation and walls of the building may be helpful in general insect and spider control.

*Exclusion*—Spitting spiders may enter under doors or through cracks around windows and may enter living spaces from the crawlspace or basement by following wires and pipes. Sealing as many cracks as possible in exterior walls and around plumbing leading from the substructure may be beneficial. Doors and windows should be equipped with weatherstripping, and soffit vents, foundation vents, and roof gable vents should be equipped with tight-fitting screens.

***Spider Removal***—Because spitting spiders are beneficial creatures, preserving them by capturing a single spider found indoors and then releasing it outside is preferred. A cup can be placed over the spider and a piece of paper slid underneath to entrap the spider. The cup can then be taken outside to release the spider.

***Insecticide Applications***—Nonchemical control strategies provide the best long term results for controlling this spider. In situations where numerous spitting spiders are present or where they are seen repeatedly, insecticide applications may be required to provide the desired results for the customer. The amount and type of treatments needed depends on the situation.

*Crack and Crevice and Void Treatment*—Cracks behind baseboards and around window and door frames may be treated using a residual dust. With larger numbers of spiders being seen, removing the outlet and switch plates and treating the wall voids behind with a residual dust may be helpful. Cracks and holes in the exterior should also be treated with dust insecticide, including weep holes.

# GROUND SPIDERS

## *Family Gnaphosidae*

**Key Identifying Characters.** Ground spiders have elongate abdomens like many hunting spiders, but most of them are uniformly colored in drab colors of gray, tan, brown, or black although there are some very spectacular species with striking black and tan coloration. The diagnostic identification feature for this family is their spinnerets. Most spiders have short, conical spinnerets at the posterior of their abdomens that are composed of two segments. However, ground spiders have a pair of relatively

Ground spider.

long, tube-like or cylindrical spinnerets that are comprised of one segment (Figure 1). Whereas other spiders' spinnerets narrow down as they reach the end, ground spider spinnerets are the same diameter the entire length. In addition, it is often the case that when looking at the spider from above, you can see the spinnerets sticking out from underneath like two exhaust pipes from a car. In most spiders with conical spinnerets, the spinnerets are tucked under the body and not visible from above. The eyes are typically equal in size with some being elliptical in shape (Figure 2, page 144).

The parson spider (*Herpyllus ecclesiasticus* found east of the Mississippi River, *H. propinquus* found west of the Mississippi River) is a very common spider in homes (Figure 3, page 144). Its body is typically dark gray with an elongate copper to tan marking on the anterior dorsal portion of the abdomen where the marking often ends in a slightly forked pattern. The mouse spider (*Scotophaeus blackwalli*) is a non-native species that is found in the Gulf Coast area and all along the Pacific Coast. It has a dusky grey coloration similar to the coat of a mouse. These are frequently found in homes. They also should NOT be

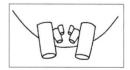

Figure 1: The tube-like spinnerets are diagnostic for the ground spider family.

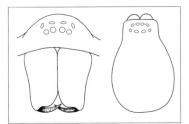

Figure 2: The face of a ground spider. Eyes are sometimes elliptical.

confused with the mouse spider from Australia (*Missulena*) which is toxic, much larger, and a much more primitive spider, more closely related to tarantulas than to the common spiders. In the western United States, a pale, non-native spider with no common name (*Urozelotes rusticus*) can be highly abundant inside homes.

Ground spiders range from 2 mm to about 12 mm in body length with 24 genera and 255 species in North America.

**Biology in General.** Ground spiders are active hunters, mostly nocturnal, and move around through vegetation, leaf litter, and grass. They make a retreat in which they rest and lay their egg sacs. Most species are found in the western and southwestern United States.

One genus, *Micaria*, has many species, is diurnally active, and mimics ants to the point where it will have white lines indenting on its abdomen, giving the impression that the spider actually has three body parts like an ant and not a spider. Some *Micaria* species have actual constrictions to mimic an ant's body more closely. These spiders also have shiny abdomens like ants in contrast to most of the other ground spiders, which have matte or drably colored body coloration. To finish off the mimicry, they may hold one pair of legs in front of them as they walk and wiggle them, making these legs look like the antennae of ants.

**Biology in Regard to Pest Control.** Although most of the ground spiders encountered by pest professionals are found in natural areas away from humans, several species are synanthropic.

**Toxicity.** Although bites have been recorded from ground spiders, almost all are minor and self-healing.

Figure 3: The parson spider is commonly found inside homes throughout North America.

**Inspection Tips.** Ground spiders are primarily outdoor spiders which occasionally enter buildings when following or hunting prey. Most of the inspection time will need to be directed outdoors. Inspections under items such as rocks, landscape timbers, and piles of debris may reveal the sac-like retreats used by these spiders during the day.

Inside, a detailed inspection may not reveal much because only one or two spiders are generally involved when ground spiders have been seen indoors. Monitoring traps placed along walls in areas where spiders have been seen are necessary to capture a specimen for identification. A nonresidual insecticide can be used to flush out spiders that might be hiding in cracks behind baseboards or door frames. Monitoring traps, however, are mostly more useful in finding ground spiders than an indoor inspection with a flashlight and a flushing agent.

**Key Management Strategies.** The best approach to managing ground spiders inside buildings is to exclude them from entering together with removal of harborages the spiders can use to locate their protective retreats. Steps taken to reduce the populations of insects outdoors that could serve as their prey will also be helpful.

*Exclusion*—Sealing cracks around windows and doors and other cracks in the exterior walls is critical in preventing spiders from entering. Doors and windows should be equipped with complete weatherstripping, and soffit vents, foundation vents, and roof gable vents should be equipped with tight-fitting screens. Exclusion is typically a key control measure for dealing with ground spiders.

*Sanitation*—Debris, rocks, leaf litter, and other items which could harbor ground spiders or their potential prey should be removed from around the perimeter of the structure. For general insect and spider control, ivy or other ground covering vegetation should be avoided next to buildings, but at the very least it should be cut away from the foundation. Branches of trees and shrubs should be cut back from touching walls and the roof.

*Exterior Lighting*—Metal halide lights on the outside of structures attract insects which, in turn, attract spiders, including ground spiders. Changing to sodium vapor lights on commercial buildings attracts fewer insects and thereby reduces spider populations. Installing yellow "bug" light bulbs in the outdoor fixtures of homes will also attract fewer insects.

*Insecticide Applications*—Nonchemical control measures are usually sufficient for relieving or preventing infestations of ground spiders.

If numerous ground spiders are entering a building, crack and crevice, spot, and perimeter treatments may be necessary.

*Directed Contact Treatment*—Individual spiders that are found can be killed by the direct application of a nonresidual aerosol insecticide.

*Crack & Crevice and Void Treatment*—Indoor ground spider retreats are likely to be located in cracks at or near the baseboard level. A light application of a residual dust insecticide into these cracks should control spiders that may be harboring there. Exterior cracks and holes should also be treated with dust insecticide, especially around windows and doors.

*Spot Treatments*—When numerous ground spiders have been seen in a room, the application of spot treatments using a residual liquid insecticide (WP, SC, or CS) to floor/wall junctures in that room can be helpful. These treatments are best made to areas under or behind furniture or other items because spiders are more likely to spend more time in protected sites like these.

Monitoring traps should also be utilized in cases where spot treatments are necessary. Remember, each spider caught on a trap is one fewer the customer might see.

*Perimeter Treatments*—When ground spiders are a particular problem in one building, a perimeter treatment with a WP, SC, or CS formulation may be beneficial. This treatment, by itself, is not likely to provide good results unless other steps such as sanitation and exclusion are completed. The application should be made to the foundation and up to several feet away from the building. Landscaped areas with heavy ground cover vegetation should receive close attention.

# SAC SPIDERS

## *Families Clubionidae, Corinnidae, and Miturgidae*

**Key Identifying Characters.** Sac spiders are generic-looking ground-running, hunting spiders with few conspicuous features that are diagnostic for their identification. Because they are so generic, many genera were dumped into the large family Clubionidae as a default decision, that is, they didn't know where else to assign them. However, recent taxonomic research has been moving genera out of the classic label of Clubionidae to other families such that where there used to be over 20 genera in the family, now only two genera are left. Nonetheless, for the pest professional, all of these families can be treated as one group.

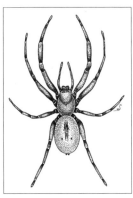

Yellow sac spider.

As one can imagine, all these spiders are rather similar looking. They have elongate abdomens and average-sized legs in length and thickness. The major identification for this group is that they have short, conical, 2-segmented spinnerets (Figure 1). In addition, most of them have claw tufts at the tip of the tarsi in which their two tarsal claws reside. The colorations are typically unremarkable monochromatic tans, browns, and grays for the cephalothorax and abdomen. In some, the skin of the abdomen is so thin on the dorsal surface near the point where it connects to the cephalothorax that it looks like a slightly different color. This is the cardiac region and, sometimes, the heart may be visible pumping fluids through the body. Many of them have widely spaced eyes (Figure 2, page 148).

The most common spiders of pest concern in these genera would be the yellow sac spiders of the genus *Cheiracanthium*. Both are synanthropic spiders with *C. mildei* being very common in homes and *C. inclusum* being

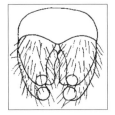

Figure 1: The sac spiders have conical spinnerets.

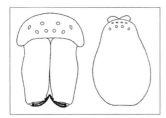

Figure 2: The eyes of many sac spiders are small and widely spaced.

less common but frequently found in agricultural ecosystems. Before 1950, *C. mildei* was not common in North America but since has spread across the continent; it is a European immigrant. *C. inclusum* is widespread in crops but is probably also another non-native as there are about 160 species of *Cheiracanthium* worldwide; it doesn't make sense that there would be only one native species in North America with all these species elsewhere.

The other major sac spiders found in homes are of the genus *Trachelas*. They have a dark magenta or cinnamon colored cephalothorax with a tan abdomen. In the eastern United States, *T. tranquillus* is a common species while *T. pacificus* is common in the west.

A variety of other species occur, some with stark, contrasting coloration that occasionally crawl into homes, but they are not major contributors to the pest control world. Overall, for these families, sac spiders range from 4 mm to about 12 mm in body length when mature with 22 genera and 197 species in North America.

**Biology in General.** Sac spiders are typically nocturnal hunters that spin a silken retreat in the shape of a sac and hide in there during the day. They emerge at night to hunt. They often make their retreat within a leaf by folding it over for greater protection. They also make their sacs under rocks, under loose bark, and in leaf litter. In homes, *C. mildei* constructs sacs in corners, often near the ceiling, under baseboards, within furniture, and behind pictures on the wall.

**Biology in Regard to Pest Control.** These spiders can be highly synanthropic although they do not typically breed inside a home as they do most of their hunting outside. They make sacs in protective places between walls and other items such as bookshelves, artwork, mirrors, etc. They are very adept at squeezing through the smallest of cracks and crevices so there is little hope of physically excluding them from a home. Because of their preference for squeezing into small spaces, yellow sac spiders received nationwide media coverage in the winter of 2011 because they crawled into and clogged the fuel lines of cars.

The manufacturer ended up recalling 65,000 vehicles for retrofitting.

**Toxicity.** If you look at older literature, it will state that yellow sac spiders can produce a toxic bite similar to that of a mild brown recluse bite. A recent study overturned this notion, showing that in verified bites, no wound developed these symptoms. This matter is covered in more detail in the *Health Aspects of Spiders* chapter (pages 39-56). Yellow sac spiders do indeed bite, it is painful like a bee sting but the symptoms are mild and self-healing without great consequence. There are also recorded bites from *Trachelas* spiders which are likewise with minor symptoms and no lasting detriment.

**Inspection Tips.** Most sac spiders live outside and rarely venture inside. Sac spiders most commonly enter buildings in the fall when the temperature cools and outside sources of prey diminish. If one of the darker-colored species is found inside a building, a detailed inspection and control efforts are most likely not needed. The yellow sac spider, however, can be found inside throughout the year and control efforts are needed as it is known to bite people.

The yellow sac spider often enters through exterior cracks and holes, and some may live inside wall voids preying on insects that have also taken up residence there. These spiders then may enter the living areas of the home through cracks around baseboards, plumbing, light fixtures, vents, etc.

This spider is generally easy to find during the daytime because its silk retreats are normally located in upper corners by the ceiling and along the ceiling/wall juncture. Gently poking the retreat with a pencil or a similar object will cause the spider to emerge where it can be captured and identified. When a small, yellowish green spider is chased out of such a retreat, it is likely that it will be a yellow sac spider. Be sure to inspect all rooms in the structure including those in a finished basement and the garage. The yellow sac spider's retreats will also be found in crawlspaces and unfinished basements along the corners formed by all of the floor joists, subflooring, box headers, and the other wooden members present in the substructure (Figure 4, page 150). Needless to say, inspections in these areas may take some time but will be necessary if the home has numerous spiders in the rest of the structure.

Outside, sac spiders, including the yellow sac spider, will be found in their silk retreats under items lying or piled on the ground such as

Figure 4: Yellow sac spiders make their sacs in very tight quarters. The two spiders in their sacs are in the groove of a sliding glass door after the door was opened.

stones, logs, landscape timbers, firewood piles, piles of lumber, and other debris. It can also be found living in outbuildings such as garages, sheds, and dog houses.

**Key Management Strategies.** A detailed control program will be needed only in those cases where the two species of the yellow sac spider are involved indoors. If other species of sac spiders are found inside a building, treatment of the exterior as described below will likely produce satisfactory results.

*Contributing Conditions*—The most important condition to address in reducing the threat of sac spider infestations is the removal of potential harborages near the foundation of the building. Firewood piles should be moved as far from a home as possible, stored off the ground, and covered with plastic to keep it dry. All items lying on the ground, such as piles of lumber and other debris, stones, and boards should be removed from the property. Heavy vegetation, such as ivy and other ground covers, as well as branches of trees and shrubs should be cut away from the building. Completing these tasks also reduces the potential harborages for insects serving as the spiders' food source.

*Exclusion*—Any crack in the exterior of a building through which spiders could enter should be sealed after they have first been treated with a residual insecticide as described below. All vents, windows, and doors should have no cracks around them and should be equipped with tight-fitting screens. It is also important to seal cracks around pipes, wires, and cables leading up into the structure from the crawlspace or basement.

Buildings with brick veneer often have weep holes to allow moisture to exit from behind the veneer. These weep holes cannot be sealed, but small pieces of screening or Xcluder™ fabric can be inserted into the weep holes to prevent spiders and other pests from entering.

*Sanitation*—The silk retreats as well as the spiders within them should be removed. The sacs may need to be scraped out of corners with a putty knife. Keeping a vacuum device handy allows quick removal of spiders and their retreats and a corner attachment for the vacuum will be useful. New activity in the future is easier to detect if webbing is removed. Removing potential outdoor harborages for spiders and sealing exterior cracks as discussed above are both part of a good sanitation program for sac spiders.

*Insecticide Applications*—Inside, removal of sac spiders in their silken retreats by vacuuming will generally get most of the spiders. If a vacuum device is not available, treatment of spiders found using a directed contact treatments can be applied. The use of crack and crevice treatments and spot treatments with residual liquid insecticides may also be helpful.

*Crack and Void Treatments*—Cracks inside where sac spiders could enter the living areas of the building from inside walls should be treated with a residual dust insecticide. Cracks around plumbing pipes, light fixtures, and air vents are examples of sites possibly requiring treatment. Sealing the cracks after treatment is recommended.

If a severe infestation of yellow sac spiders is encountered, removal of all of the plates covering electric outlets and switches and treatment of the wall void behind with a dust insecticide may be required. This treatment, however, should be rarely necessary for most sac spider infestations.

In crawlspaces and basements, application of dust insecticides into the cracks under sill plates and into the voids of foundation walls may be needed. This may be particularly necessary for older buildings with stone or brick foundations. Dust insecticides also can be applied to the sill plate areas of crawlspaces.

*Spot Treatments*—The application of liquid residual WP, SC, or CS formulation to upper corners of rooms following removal of spiders can be helpful in some cases involving these spiders.

In crawlspaces and basements, spot treatments can be applied to the sill plate/box header area at the top of foundation walls.

*Exterior Treatments*—The most important control measure for outside areas is the removal of as many potential harborages as possible as discussed earlier in this section. This automatically results in a reduction of spiders and the insects they prey upon near the building.

Cracks in the exterior walls of the building should be treated with a residual dust insecticide and then sealed to prevent spiders from entering in the future. Use of a large hand duster or cordless electric duster may be beneficial to apply dust through weep holes and kill spiders that might be harboring behind the brick veneer.

Spot treatments with a WP, SC, or CS insecticide around doors and windows may be useful in preventing new spiders from entering.

Perimeter treatment to foundations and eaves may have limited effectiveness for sac spiders and probably may only be necessary for severe infestations with a high possibility for reinfestation from outside sources.

# LYNX SPIDERS

## *Family Oxyopidae*

**Key Identifying Characters.** The diagnostic feature that identifies these spiders is that six of their eyes are arranged in a hexagon on the front portion of the carapace with two smaller eyes located below closer to the chelicerae (Figure 1). The spiders also have relatively thin legs with highly conspicuous, long dark spines sticking out almost perpendicularly to the length of the leg. The green lynx spider (*Peucetia viridans*) is probably the most commonly encountered lynx spider in North America, restricted to the southern portion of the United States (Figure 2, page 154). It is bright apple-green in color and one of the few spiders that actually can be accurately recognized by homeowners. Other lynx spider species of the genus *Oxyopes* are brown and tan and not as brilliantly colored (Figure 3, page 155). They range from 5 mm to about 17 mm in body length when mature with three genera and 18 species found in North America.

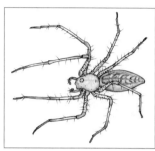

Lynx spider.

**Biology in General.** Lynx spiders are agile, day-active hunters with excellent eyesight. Their vision is on the same level as wolf spiders. They are typically found in vegetation where they run from leaf to leaf and capture prey.

Although the green lynx spider is mostly green, it has a pattern on its dorsal abdomen, which is composed of various degrees of white, red, brown, and purple. The spider can change its color over a period of about 10 days to better match its background. On the same day, one green lynx spider was on

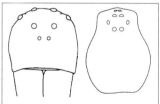

Figure 1: Six of the lynx spider's eight eyes are arranged in a hexagon.

a green plant with small white flowers and was mostly green and white (see color photo section of the book) while just 10 feet away, another green lynx spider was colored in more browns and purples similar to the dying sage plant that it was on. It is unlikely that weather or time of year had much influence over coloration.

The egg sac of a green lynx spider is covered with bumps. The spiderlings that emerge are orange.

**Biology in Regard to Pest Control.** These spiders typically are not associated with homes so there is little to be said in regard to pest control. However, because lynx spiders do roam, occasionally one will enter a house or be carried inside on a plant. More often, the very conspicuous green lynx spider, which can get rather large (up to 17 mm for a female about to lay eggs), is found on the outside of a house, thus causing concern when people see such a strikingly colored spider. Despite its beauty, the homeowner typically still wants the spider eliminated.

Figure 2: The green lynx spider is usually bright green in coloration and has lots of spines sticking out almost perpendicularly from its legs.

**Toxicity.** Because of their size, green lynx spiders are capable of biting, but the bite leaves only a little red mark and is self-healing. However, green lynx spiders are capable of spitting venom about a foot or so. This was discovered by an arachnologist who was studying these spiders and when she got close, she felt small drops of wetness hit her face and found little drops of fluid on her glasses. People who have received a green lynx venom shot into their eyes had mild redness and irritation for about a day before recovering.

**Inspection Tips.** Lynx spiders are rarely found inside structures. These spiders are more likely to be encountered by a pest management professional outside on plants and possibly on the exterior walls while performing services to control other pests. If a lynx spider is found inside of a building, it is likely to be associated with a plant that has been carried indoors or it has crawled in through a window or doorway.

**Key Management Strategies.** Lynx spiders are highly beneficial creatures and most likely will never need control efforts directed at them.

It is best to capture any lynx spider found inside and release it outside. Because these spiders are so quick, the best way to capture them is to place a cup over the spider and then slide a piece of paper under the cup to trap the spider inside. Release captured spiders onto a plant or on the ground in a landscaped area.

Figure 3: The smaller lynx spiders, *Oxyopes* sp., are brown but still have the conspicuous leg spines.

To prevent future invasions by this spider, cut tree and shrub branches away from the walls of the structure, keep other vegetation cut away from the foundation, and seal cracks in the exterior around windows and doorways.

No treatments should be necessary for this spider.

# WOODLOUSE SPIDER

## *Family Dysderidae*

**Key Identifying Characters.** The woodlouse spider is a non-native species in North America. It has a dark cinnamon or magenta cephalothorax and a gray to tan abdomen. It has six eyes, in two tightly clustered trios near the front of the cephalothorax (Figure 1). It also has very long fangs which it will brandish toward a threat if cornered. It ranges from 9 mm to about 15 mm in body length when mature. Its coloration is very similar to the sac spider (genus *Trachelas*) for which it is sometimes mistaken, but the woodlouse spider has six eyes in tight clumpings whereas the sac spiders have eight eyes set very far apart.

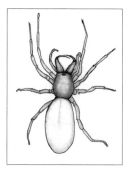

Woodlouse spider.

However, these two spiders are rather unique in coloration so once you are able to separate these by their eye patterns, it should be easy to correctly identify them. Only one genus and one species is found (*Dysdera crocata*) (Figure 2, page 158) in North America although there are several hundred species in the Eastern Hemisphere.

**Biology in General.** The woodlouse spider is very common throughout North America and can be found inside homes. It only recently became established in the Pacific Northwest. It can be found under rocks, loose bark, rubbish, and clutter. In drier climates, it seems to be found only under objects like trashcans where the humidity is higher and protection from heat and arid conditions occurs.

Their long fangs are supposedly designed for the killing of woodlice (pillbugs) where they grab the rolled up crustacean and are able to insert their fangs into

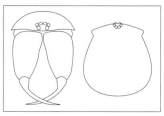

Figure 1: The woodlouse spider has six eyes which are clumped together in the front of its cephalothorax.

the woodlouse's body to kill it. They also will take other prey as well, like a normal spider.

**Biology in Regard to Pest Control.** These are occasionally house invaders, usually entering in search of prey.

**Toxicity.** A few recorded bites of the woodlouse spider have been made with reports of only mild transient pain after the bite. Much of this pain is probably due to punctures made by the large fangs.

Figure 2: The woodlouse spider has long fangs and a dark magenta colored cephalothorax.

**Inspection Tips.** Woodlouse spiders are primarily outdoor spiders which occasionally enter buildings when following or hunting prey. Most of the inspection time should be directed outdoors. Inspections under items such as rocks, landscape timbers, and piles of debris may reveal these spiders during the day.

Inside, a detailed inspection may not reveal much because only one or two spiders are generally involved when this spider has been seen indoors. Monitoring traps placed along walls in areas where spiders have been seen may be necessary to capture a specimen for identification.

**Key Management Strategies.** The best approach to managing woodlouse spiders inside buildings is to exclude them from entering together with removal of harborages outside where they might hide.

*Exclusion*—Sealing cracks around windows and doors and other cracks in the exterior walls is critical in preventing spiders from entering. Doors and windows should be equipped with tight-fitting weatherstripping.

*Sanitation*—Debris, rocks, leaf litter, and other items which could harbor woodlouse spiders or pillbugs and sowbugs, should be removed from around the perimeter of the structure. Ivy or other ground covering vegetation and thick mulch layers should be avoided next to buildings as they serve as excellent harborages for pillbugs and sowbugs.

***Insecticide Applications***—Nonchemical control measures are usually sufficient for relieving or preventing infestations of these spiders. If numerous woodlouse spiders are found to be entering a building, crack and crevice or spot treatments may be necessary.

*Crack and Crevice and Void Treatment*—Exterior cracks and holes should be treated with dust insecticide, especially around windows and doors.

*Spot Treatments*—When numerous woodlouse spiders have been seen in a room, the application of spot treatments using a residual liquid insecticide (WP, SC, or CS) to floor/wall junctures in that room may be helpful. These treatments are best made to areas under or behind furniture or other items because spiders are more likely to spend more time in protected sites like these.

Monitoring traps should also be utilized in cases where spot treatments are necessary. Remember, each spider caught on a trap is one fewer the customer might see.

# CRAB SPIDERS

## *Family Thomisidae*

**Key Identifying Characters.** As the name suggests, crab spiders are so called because of their resemblance to crabs. The overall body form is fairly diagnostic for the family. Their legs stick out to the side and are rotated 90° backwards so that the dorsal surface of the leg segments actually face laterally. The first two pairs of legs are longer than the hind two pair of legs and the spider can move like a crab sidewise and backwards.

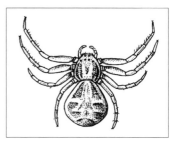

Crab spider.

They also have a smattering of spines sparsely covering their bodies and are either various shades of brown resembling tree bark, or they can be bright white or yellow, sometimes with pink stripes. Their eyes are typically small and all about the same size and are often on small turrets (Figure 1). The abdomen is flattened, being much thinner than long except in a female about to lay an egg sac. Often, the abdomen is much wider at the posterior end or angular in form around the edges. Males are much smaller than females and have proportionally much longer legs than the females. One of the most common yellow crab spiders is *Misumena vatia*; many species of brown crab spiders occur, but the most common genus is *Xysticus* with *X. funestus* being one of the most widespread species. Crab spiders range from 3 mm to about 11 mm in body length when mature with nine genera and 130 species in North America.

**Biology in General.** Crab spiders are found in habitats matching their coloration. The brown species can be collected under tree bark and in leaf litter, or possibly running through grass (Figure 2, page 162). The lighter colored

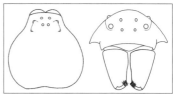

Figure 1: The crab spider has eight small eyes on raised prominences.

PCT Field Guide **161**

Figure 2: The crab spider has a wide, flat abdomen with its four front legs being longer than the four hind legs.

species rest in the heads of flowers where they await pollinators, which they typically grab behind the head with their fangs. Many honey bees and nectar-feeding flies have their lives ended on flower blossoms. In addition, the white and yellow species are able to change their color over a period of a few days so that they match the influorescence more closely. So a white spider can become a yellow spider and vice versa, given ample time.

**Biology in Regard to Pest Control.** These spiders should not be of great concern to the pest control industry because they rarely enter homes. Most interaction with humans will be from people finding them in their gardens on flowers. The darker species may occasionally be found on humans who hike, walk through brush, or are out golfing. Because of their flattened bodies and crab-like appearance, they can be mistaken for ticks.

**Toxicity.** A few recorded crab spider bites have been reported with minor, self-healing effects.

**Inspection Tips.** Crab spiders are rarely found inside structures preferring to stay outside. Occasionally, however, one might be found inside usually along a baseboard, in a window sill, or captured on a monitoring trap located near a doorway. Sometimes, a homeowner will notice a crab spider on a plant or fresh cut flowers carried inside from the garden or patio. A pest management professional is more likely to see crab spiders outside on walls or in cracks around windows while performing services to control other pests.

If a crab spider is found inside, it is generally not necessary to expend much effort to inspect inside for additional spiders—they are not likely to be found. If several crab spiders were to be found inside a building, they are likely to be associated with a plant that has been carried indoors or they have crawled inside under or around a window or doorway. This situation will rarely be encountered.

**Key Management Strategies.** Crab spiders are only a danger to bite if they are picked up and held. Otherwise, they are highly beneficial creatures that should not need control efforts directed at them.

Individual spiders can be removed when found inside. Ideally, it is best to capture the spider and release it outside. This can be done by using forceps to pick up the spider and place it into a container or to entice the spider to crawl onto a piece of paper so it can be carried outside. Release the spiders onto a plant or on the ground in a landscaped area.

To prevent future invasions by this spider, cut tree and shrub branches away from the walls of the structure, keep other vegetation cut away from the foundation, and seal cracks in the exterior around windows and doorways.

The only treatments that might be necessary for this spider are:

• The application of a residual insecticide into cracks around doors and windows before they are sealed.

• The application of a spot treatment to surfaces around windows and doorways outside.

# RECLUSE OR VIOLIN SPIDERS

## *Family Sicariidae*
## *Genus Loxosceles*

**Key Identifying Characters.** The most definitive way to identify a recluse spider is to look at the eye pattern. Whereas most spiders have eight eyes usually in two rows of four, recluse spiders have six eyes with a pair in front and a pair on either side, along a U-shaped line with a space between each pair (Figures 1 and 2, page 166). This eye pattern is also found in spitting spiders

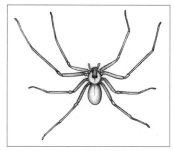

Brown recluse spider.

(*Scytodes* species, page 139) but the coloration of spitting spiders is sufficient to exclude consideration as a recluse.

Historically, pest professionals relied on the presence of a violin-shaped marking on the dorsal surface of the cephalothorax to identify a brown recluse spider from other spiders. Although brown recluse spiders (*Loxosceles reclusa*) do indeed often have a very distinct dark brown violin (Figures 2 and 3, page 166), this marking is not conspicuous in the younger spiderlings or in a brown recluse that has recently molted (see Figure 9 in the *Basic Spider Biology* chapter). In addition, several of the southwestern *Loxosceles* species have no pigmentation in the violin area (Figure 4, page 166) and, therefore, they are often misidentified as common harmless, tan spiders of which there are many in the deserts. Also, the female Chilean

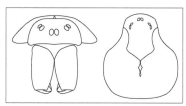

Figure 1: Recluse spiders can be identified by the arrangement of their six eyes on the carapace. Often, a distinct to faint violin-shaped marking is present on the carapace.

Figure 2: The violin pattern and the six eyes are readily distinguishable on this brown recluse spider.

recluse (*Loxosceles laeta*) has a reddish brown cephalothorax so the violin marking is often not distinguishable from the rest of the carapace (Figure 5). However, the most important reason to not use the violin marking on a recluse spider as the definitive identification mark is that, because of the infamy of the brown recluse, people are extremely creative in misinterpreting any kind of marking whatsoever on any portion of a spider's body whatsoever as a violin marking. The number of spiders that are misidentified as brown recluses this way is mind-numbing. Non-arachnologists are also confused by images from the Internet because recluse abdominal color can vary from a light cream color to almost black depending on what they feed upon, thus providing tremendous variation.

In addition to the violin marking and eye pattern, other aspects of the physical features of a brown recluse can be used to distinguish that a spider IS NOT a recluse spider. Table 1 (page 167) may help in convincing clients that they are wrong despite their adamant attitude.

Recluse spiders have a characteristic way that they sit when at rest with the front three pairs of legs crooked and facing forward with the fourth pair of legs behind, sometimes straight behind (Figure 6, page 167). This posture actually is what gave these spiders their scien-

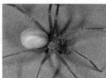

Figure 3: The brown recluse often has a distinct violin pattern on its cephalothorax but it is better to use the pattern of six eyes to confirm identity.

Figure 4: The desert recluse has no pigment in the violin area and is often mistaken for a harmless spider.

Figure 5: The female Chilean recluse has a dark cephalothorax so its violin pattern is not easily discernible. It lives in Los Angeles County and has been found only once or twice elsewhere in the country.

Table 1. Identifying characters of recluse spiders compared to other spiders.

| Non-recluse spiders can have: | Brown recluse spiders have: |
|---|---|
| Legs with thick spines | Legs with fine, flat-laying hairs; no spines |
| Legs with stripes, rings, or spots | Legs are uniformly colored |
| Legs are dark brown or black | Legs are tan, sometimes first legs orange |
| Abdomen has more than one color | Abdomen uniformly colored |
| Body length can be more than ⅜" | Body length no more than ⅜" |

tific name. "*Loxosceles*" means "slanted legs."

Recluse spiders range from 6 mm to 13 mm in body length when mature. One genus and 13 species are found in North America, two species of which are limited localized infestations of non-native species.

**Distribution.** The brown recluse spider (*Loxosceles reclusa*) is the most infamous of the *Loxosceles* spiders in North America. However, despite the fact that people throughout the continent believe

Figure 6: Recluse spiders have a characteristic positioning with their front six legs crooked and pointing forward with the fourth pair of legs pointing backwards.

that brown recluses are everywhere, including such improbable localities like Canada and Alaska, this species is limited to the central Midwest (Figure 7, page 168) from southeastern Nebraska to Texas and east to southwestern Ohio and northwestern Georgia. As one gets to the edges of the distribution, it becomes much more difficult to find them. For example, a study in Georgia could only locate about 100 records of brown recluse spiders collected in the state both historically and during a 5-year study. To find 100 brown recluses in Kansas or Missouri would probably require about an hour or two in a heavily infested structure. Also, from opinions of arachnologists and the data from a study that was recently started for the Gulf Coast states, brown recluses are very rare near the coast and become more common as one heads inland. For example, a board certified entomologist from coastal Houston, Texas

Figure 7: Distribution of recluse spider species in the U.S.

says that if he wants to find brown recluses, he has to go inland about 100 miles. Historical records from Louisiana, Mississippi, and Alabama show them to be common in the northern portions of the state and sporadic in the southern portions. For the moment, not enough data exists to make definitive statements about these areas; however, when enough solid data is accumulated, the distribution of the brown recluse will probably be moved inland about 100 miles from the coast.

Five other species are found in the southwestern deserts with widespread distribution, but they are mostly associated with natural desert or homes surrounded by natural vegetation. These include the desert recluse (*L. deserta*), Arizona recluse (*L. arizonica*), apache recluse (*L. apachea*), Texas recluse (*L. devia*), and the Big Bend recluse (*L. blanda*). The other five native species are found in isolated canyons, some only from that one locale. The Mediterranean

Figure 8: The Mediterranean recluse looks very similar to the brown recluse. It requires a microscope and much experience to separate the two species.

**168** Recluse or Violin Spiders

recluse (*L. rufescens*) is a worldwide tramp species and shows up sporadically in isolated buildings around the world (Figure 8, page 168). Several such buildings exist in the United States, but almost every one is a unique infestation. As is typical of *Loxosceles* infestations in new areas, usually only that one building is infested or nearby buildings if connected with conduits and piping. Once one infestation is located, it may require an extended search before a second infested building is found nearby. The Chilean recluse (*L. laeta*) has very isolated establishment in the United States. A population has been surviving in urban Los Angeles County since the 1930s, however, infested buildings are few and far between. The spider has not spread very far in the 80 years that it has been there and it is not considered a public health issue by the Los Angeles Department of Health. No verified bites or even complaints of bites have emanated from Los Angeles County. Another infestation was present in the the Museum of Comparative Zoology at Harvard University, however, it is unknown if that populations still exists.

Figure 9: Recluses often feed on their prey by sucking the nutrients out through a leg as shown in the picture here.

**Biology in General.** In nature, recluse spiders are typically found in dry places, under loose tree bark, and under rocks and overhanging outcroppings. They are rather primitive spiders and have slow metabolisms. Whereas many more evolutionarily advanced spiders will process prey in an hour or so, recluses may extract nutrients from their conquest over a 24-hour period. They are sometimes comical in their feeding where they will sit for hours with just a leg or an antenna in their mouthparts, sucking up nutrients similar to people using a straw in a drink (Figure 9).

Recluses are highly limited in their ability to disperse. In the introduction chapter of this book, it was mentioned how spiders disperse over great distances by ballooning behavior during warm days. However, recluse spiders belong to a primitive grouping called haplogynes, which do not balloon. Therefore, recluses can only disperse by walking or having people move them around.

As an example of the lack of dispersal in recluse spiders, publications from Georgia described the first finding of brown recluse spiders in a natural setting, which was under the bark of a tree. Seven years later, using the field notes of the collector, this same tree was relocated. It has fallen down with age yet brown recluses were still collected under the loose bark of this tree. The scientists searched all around the area for about 1/2 mile in other locations that would be prime recluse habitat (an old collapsed building, a woodpile) and found no recluses. So even though these spiders had years to spread, they were only found at that one tree. In addition, an arachnologist did a yearlong study in his backyard in Lawrence, Kansas where he looked at the spiders that crawled into cardboard rolls every month. He collected more than 700 spiders in one year and not one was a brown recluse, despite the fact that recluses are very common in Kansas. This begs the question of how recluses do actually move around without human help, and currently we have no good answer to this question.

**Biology in Regard to Pest Control.** Of the species in North America, the brown recluse is synanthropic making it a significant pest species where it occurs. It populates homes and other buildings, sometimes in large numbers. It is not uncommon to have hundreds to thousands of brown recluses inside a single home and often without a medical incident (bites). In nature, they squeeze themselves under rocks and bark; in homes, they squeeze themselves behind bookcases, in holes and cracks in cabinetry, under tar paper and insulation, between wood timbers, etc. Three recent laboratory and field studies showed that recluse spiders prefer crevices that are vertical in orientation rather than horizontal, have angular interiors rather than rounded, and have silk from previous recluse inhabitants rather than clean refuges.

In recent years, research was put forth that brown recluse spiders preferred to scavenge rather than eat live prey and that they could eat pesticide-killed prey without picking up a lethal dose themselves. The extrapolations from this research suggested that spraying a house for pest control could increase brown recluse populations inside homes. These counterintuitive notions caused confusion for homeowners and controversy in the pest control industry. Since then, additional research has shown that scavenging behavior is flexible and that recluses preferred to capture live prey instead when factors such as prey size, predator hunger levels, and length of time that prey have been dead were varied.

Another laboratory study showed that many spiders from a wide spectrum of spider families likewise scavenge prey if given the chance so there is nothing special about brown recluses in regard to scavenging. At that point, assuming equal susceptibility to pesticides, whatever happens to brown recluses when a house is sprayed should also happen to other spiders. The impact of a pesticide application in a home and its effects on spider populations has yet to be fully answered, but it seems prudent to be cautious about conclusions regarding spider scavenging until actual field tests can be run in homes. In a South American study, an attic was sprayed for *Loxosceles* spiders with lethal effect being from residual not contact spray and, indeed, many recluse spiders were killed by the application.

**Toxicity.** Recluse spiders are well known for having venom that is toxic to humans. They can cause severe skin injuries and, in rare cases, death; however, most recluse bites are self-healing and do not leave significant scars. It is only the rare bite that results in rotting flesh (also known as necrotic wounds). All species of recluse spiders examined so far contain the venom component that causes necrosis, so they all should be considered medically important. See the chapter on *Health Aspects of Spiders* for more information.

**Inspection Tips.** Because brown recluse spiders are so secretive, it can be difficult to find all of their harborages in order to treat them. One of the key signs to look for is the irregularly shaped, whitish-gray webs which may be located inside cracks and boxes, under insulation in attics and crawlspaces, and under items on the ground outside and those stored inside. In heavily infested buildings, these webs may be found in corners in between the rafters and ceiling joists in attics and along the sill plate and subfloor areas in crawlspaces and basements.

Outside, the brown recluse is almost always associated with items lying or piled on the ground. It is rarely found living strictly in vegetation, even heavy vegetation such as ivy. This spider will be found harboring under stones, logs, landscape timbers, firewood piles, piles of lumber, and other debris. It can also be found living in tree holes, under loose bark, and inside and under outbuildings such as sheds and dog houses.

Before beginning the intensive control efforts needed to eliminate infestations of this spider, *it is a good idea to confirm that brown recluse spiders are actually present.* The homeowner may have a specimen

Figure 10: This typically spread-legged shed skin is evidence of a historical presence of recluse spiders in a structure.

to be identified, which will quickly confirm whether these spiders are present. If the specimen is that of another spider, an inspection may still be worthwhile to put the homeowner's mind at ease. In some of these situations, brown recluse spiders are actually found, but in most cases they are not.

One telltale sign of recluse spider presence in the absence of spiders is that recluse spiders leave a diagnostic shed skin behind (Figure 10). Because recluse spiders stretch out their legs in a star-like position prior to molting and because the skin does not shrivel afterward, the remnant skin can be found where recluse spiders have been living. Typically, molting is done on a vertical surface with the head pointing upward, however, they are capable of molting on a horizontal surface, such as a ceiling, as long as they can drop down with gravity to pull themselves out of the skin. Also, recluses show site fidelity so they will return to the same retreat for many weeks. It is not uncommon to find several shed skins, from tiny to large, like Russian nested dolls, where the spider has shed in the same place several times.

A person in a home or business may report receiving a bite that has been diagnosed as being from a brown recluse. First, many skin wounds may be diagnosed as being caused by a brown recluse; however, such diagnoses are often frequently incorrect. See *Health Aspects of Spiders* in this book for more details. This is another reason to confirm whether brown recluse are actually present within the building.

Second, it is often difficult to determine where a confirmed spider bite actually occurred because of the delayed development of bite symptoms. It is especially difficult when a person who has been bitten claims that a brown recluse bite was received at work. Further investigation can reveal that no brown recluse spiders are actually at their workplace. In other instances, a homeowner or business owner has had no reports of bites but has seen spiders and fears that brown recluse spiders may be present. In all of these cases, the actual presence of brown recluse spiders and the extent of the infestation should be determined prior to beginning any control program.

In one case, a large plastics manufacturer had two workers who had been diagnosed with brown recluse bites and claimed they were received at work. A visual inspection revealed no brown recluse spiders, but large numbers of cellar and orb-weaver spiders were found living in corners and along structural beams throughout the warehouse. Numerous other insects were discovered which were entering through open doors in the warehouse. To confirm whether brown recluse spiders were present, over 200 monitoring traps were placed throughout the facility. The use of these sticky traps is the best method for confirming the presence of brown recluse spiders (as well as other types of hunting spiders). Brown recluse spiders actively move about along walls at night and will enter the traps. In this case, the traps remained for several nights before being checked. Wolf spiders, ground spiders, beetles, silverfish, and harvestmen were captured, but not one brown recluse spider was captured. General spider control services were rendered in this case, but the detailed efforts of a brown recluse program were not necessary.

***Inspecting Living Areas of Buildings***—Before placing monitoring traps, a flashlight inspection combined with a flushing agent (pyrethrins) can often locate brown recluse spiders that may be present. Begin in the rooms where either spiders were seen or a person believes they received a bite. Inspect under beds—especially the headboards of waterbeds—focusing on cracks under and behind the baseboard. Flushing agent can be used to flush spiders out of suspect cracks and voids. Pull out drawers in dressers and nightstands and look for spiders or their webs. Inspect the folds of curtains, especially at the top. Remove items from closets and inspect the folds of clothing stored there and inside boxes. It may be a good idea to wear gloves to avoid accidental bites. Check for cracks around vents and light fixtures in the wall and the ceiling because spiders living in the attic and crawlspace often follow air ducts and electrical wires into the living areas of the building. Follow this procedure for other living areas of the home or building.

Finding the source of an infestation can be difficult and time-consuming. In one case related by a pest control technician, a small boy had endured several skin lesions that were diagnosed by a doctor as brown recluse bites and repeated efforts had failed to find and eliminate any infesting spiders. Finally, an inspection revealed several spiders living behind the molding around the window located about one foot from the boy's bed. After killing these spiders, the molding was removed revealing a large crack leading to the outside. Numerous brown recluse spiders were found living in a fence post just outside of this window. The spiders were entering the crack under the window and at night were wandering across the curtains onto the boy's bed. Treatment of the spiders outside and in the crack, sealing the crack with expanding foam, and moving the bed away from the window's curtains prevented future bites.

If spiders are seen, it can be easier and faster to use a vacuum device to remove spiders and their webbing as they are found. Spiders can also be treated with a direct contact treatment using a nonresidual flushing agent. If the spiders are inside a box or in the folds of curtains or clothing, the spider should be knocked off the fabric or out of the box before treating it—the pyrethrins aerosol could damage the fabrics or items stored in the box.

***Inspecting Basements, Garages, Crawlspaces, and Attics***—Most of the brown recluse spiders in an infestation are generally located in areas such as the basement, garage, crawlspace, and attic. Boxes and other items stored in these locations provide clutter that is ideal for harboring both spiders and their prey. Gloves and long sleeves are recommended when inspecting these areas to avoid accidental bites. Rubber bands can also be placed over the bottom of pant legs to prevent spiders from running up under the pants when larger numbers of spiders are encountered. In one home that was severely infested, the service technician wore a bee veil while working in the attic to prevent spiders from falling off the roof sheathing into his collar. Once he started his inspection and treatment, spiders began to run everywhere, some of them dropping off the roof sheathing.

Boxes that are stored in an attic, basement, or garage should be inspected for spiders and the spiders should be removed or killed. Open a few boxes and check for spiders and if they are found, the boxes will need to be handled as discussed in the management section of this chapter.

Besides boxes, any item stored in these areas should be moved and inspected underneath and behind. Inspection under exposed insulation in the walls in basements and garages, under the subfloor in crawlspaces and basements, and in the floor of attics may also be necessary, depending on the situation.

Inspections can be time-consuming. When spiders are found, remove them with a vacuum device or treat them directly with a nonresidual flushing agent as described earlier in this section.

***Inspecting the Exterior***—As stated earlier in the biology section, brown recluse spiders generally don't travel far from their harborages; therefore, invasions from outdoor harborages are not likely to occur. Experience has shown, though, that brown recluse spiders can be found residing in exterior cracks and in piles of items and other debris next to structures.

Any exterior cracks in the exterior of the building should be treated with flushing agent to chase out spiders that might be residing there. Outbuildings such as sheds, garages, boat houses, and dog houses should be carefully inspected. The underside of decks, especially those located close to the ground, can harbor brown recluse spiders. It may be difficult to inspect under decks close to the ground, so these decks should be treated underneath.

Brown recluse spiders seem to be more common in attics with poor ventilation. This is probably due to the fact that insects are more common in poorly ventilated attics—a larger food supply should result in more spiders. While inspecting outside, check the number of soffit vents. If these vents are absent or are few in number, the customer may benefit from improving attic ventilation. In addition to making the attic less attractive to insects and spiders, the heating and cooling efficiency of the home may be increased.

**Managing Brown Recluse Spider Infestations.** Once brown recluse spiders have been confirmed in a building and the extent of the infestation determined by inspection and/or monitoring traps, control measures can be employed to—hopefully—eliminate the current infestation. Because these spiders are reclusive in nature and often harboring within inaccessible parts of a home or building, accessing every active harborage for treatment is generally impossible. This is especially true in older buildings which have a large, initial spider population. Certainly, control efforts will greatly reduce an infestation, and the customer may not see any spiders for long periods of time. In some cases, however, the spiders reoccur periodically so continued inspection, monitoring, and treatments may be necessary for some structures. In severe cases involving older or large complex buildings, fumigation of the structure has been done. Even whole-structure fumigation, however, has failed to totally eliminate brown recluse spiders for long periods in some cases.

A pest control company probably should avoid providing a guarantee of total elimination because of the difficulty of achieving this goal. Control programs should target elimination as a goal; however, it is always

Figure 11: Clothes that are infrequently worn, such as sweaters, should be stored in plastic boxes with tight-fitting lids.

possible for spider harborages to be inaccessible for treatment, thus allowing some spiders to survive. Also, no guarantee should be given that new bites will not occur. A person could receive a bite anywhere, not just in their home. A company which specializes in brown recluse spider control may develop a service agreement specifically for these spiders. This service agreement should spell out the components of the service and outline what is guaranteed and what is not.

***Practices to Avoid Possible Bites***—Homeowners who have brown recluse spiders can practice certain habits in order to minimize possible bites until the infestation is effectively controlled. Because bites generally occur when trapped against the skin inside clothing, clothing can be stored in sealed plastic bags inside drawers or inside plastic storage compartments hanging in a closet (Figure 11). Shoes can be stored inside plastic shoe boxes. Clothes that have been left on the floor, in a clothing basket, or are otherwise exposed should be shaken well before putting them on.

Beds should be moved out so they do not touch walls or curtains. Bed skirts around the box springs should be removed from beds and bedspreads that come near or touch the floor should not be used (Figure 12).

Figure 12: This picture indicates three things that need to be done to reduce chances of brown recluse spider bites in endemic areas: remove bed skirts, do not use the underside of the bed for storage, and do not leave clothes and shoes on the floor.

Such items allow spiders easier access to climb onto the bed and under the covers. Also, persons living in infested homes should get into the habit of inspecting bedding prior to climbing in to go to sleep.

***Contributing Conditions***—Long-term relief from brown recluse spiders involves correcting and/or removing the conditions contributing to an infestation, where possible. Outside, items lying on the ground, such as piles of lumber and other debris, should be removed from the property (Figure 13). Firewood should be moved as far from a home as possible, stored off the ground and covered with plastic to keep it dry. Heavy vegetation such as ivy and other ground covers should be cut at least 18 inches away from the building foundation. Branches of trees and shrubs touching the house should be cut back away from the roof and walls of the building.

Figure 13: Piles of items can harbor recluse spiders. This pile of tiles next to a daycare center contained several brown recluse spiders.

As stated above in the inspections section, improving poorly ventilated attics would likely be beneficial. Numerous soffit vents, together with ridge vents, make the best ventilating system for attics. The roof areas above small enclosed porches, carports, and breezeways also may benefit from installation of ventilation. Improved crawlspace ventilation can also be important for general insect and spider control.

***Exclusion***—It may be beneficial to have customers seal cracks in the exterior of a building through which spiders could enter after they have first been treated with a residual insecticide as described below. Sealing cracks that may harbor spiders before treating them may force the spiders inside. Foundation vents and soffit vents should be tight-fitting and be equipped with screening.

Buildings with brick veneer often have weep holes to allow moisture to exit from behind the veneer. These weep holes cannot be sealed, but small pieces of screening or Xcluder™ material can be inserted into the weep holes to prevent spiders and other pests from entering.

It is also recommended to seal cracks around pipes, wires, and cables leading up into the structure from the crawlspace or basement and down from the attic. Sealing cracks around ceiling vents and light fixtures may also be helpful to prevent spiders from entering the living space below.

***Boxes and Other Stored Items***—When spiders are found living inside stored boxes, all boxes should be inspected. This is generally the customer's responsibility, however, enterprising pest management professionals have occasionally sold their services to do this for their customer. This can be an important step in controlling the brown recluse spider. Leaving spiders untouched inside boxes allows a ready reinfestation source. Gaining the customer's cooperation in completing this task can sometimes be the most daunting task in brown recluse spider control. Persons who have been bitten or have a family member that received a bite, however, are generally more motivated to do anything it takes to prevent a future reoccurrence of a bite.

Boxes that are to be inspected should be moved from attics and basements to a garage or outside. It is easier to inspect them there and if a spider manages to crawl out of the box, it will be easier to deal with in a garage or outside. Any person moving or going through boxes may want to wear long sleeves and gloves to avoid contact with spiders. A vacuum device should be kept handy to quickly remove spiders as they are encountered. All items in a box should be removed and opened or unfolded to check for spiders.

Figure 14: Taping boxes prior to storing them in attics, basements, or garages prevents recluse spiders from establishing a hiding place inside.

After careful examination, the items can be replaced in the old, now spider-free box or into a new box or plastic tub with a tight-fitting lid. All corners and cracks of boxes should be taped to prevent spiders from re-entering the box (Figure 14). Where possible, loose items stored in an attic, garage, or basement should be inspected carefully then stored in a sealed box.

Where possible, it is best to store boxes at least 8-10 inches off the floor and the same distance away from the wall. Such storage practices allow for better inspection and treatment and provides space on which to place monitoring traps.

*Sanitation*—The use of a vacuum device during the inspection, as stated earlier, is an effective and time-saving control measure when dealing with brown recluse spiders (and other spiders). Vacuuming quickly removes spiders and their webs as they are discovered. Every spider removed is one more the customer will not come into contact with. Removal of webbing also helps during future inspections to determine new activity. A shop-vac® may kill most recluse spiders from the mere tumbling through the tube on the way to the canister.

Repacking boxes, storing items in a more orderly fashion, and removing potential outdoor harborages for spiders as discussed above are all part of a good sanitation program for brown recluse spiders.

*Monitoring*—The use of monitoring traps can be a critical part of a brown recluse spider control program. These traps are first used in determining the presence and extent of the infestation; they should then be continuously used throughout the entire program. Not only do monitoring traps alert the service technician that spiders are still present or have reinfested, but each spider trapped is one that will not come into contact with anyone in the building. In this manner, monitoring traps also serve as a control tool.

Figure 15: Sticky traps work well for detecting the areas where recluse spiders may be active.

In cases of moderate to severe infestations involving large numbers of spiders, use plenty of traps. The more traps used, the more spiders that should be trapped and removed. It is rare, but occasionally, infestations will be encountered where monitoring traps will literally be covered from end-to-end with spiders (Figure 15). This can be a little scary to find a situation with this many spiders and is one case where there may be no such thing as using too many traps.

Traps should be placed in all infested areas of a structure, usually along walls because brown recluse spiders generally crawl about using walls as a guide. Place traps along the wall behind furniture, behind the toilet, inside the sink vanity, inside and under kitchen cabinets, and in the space below a bathtub, if accessible. (If one cannot gain access under the bathtub, an access panel can be installed to allow inspection and treatment.) In infested closets, move items away from walls and place traps at the floor level and possibly one or two on the shelves.

In basements and garages, place traps along the floor behind items stored there. Traps also can be placed on exposed sill plates in basements and crawlspaces. Numerous traps may also be placed in infested attics, especially around boxes and near openings where light fixtures and vents are present in the attic floor.

Traps should be checked during every service. If only one or two spiders are captured on a trap, write the number captured on the side of the trap so during future checks, it will be easy to determine if new spiders have been captured. Traps capturing numerous spiders should be replaced with new traps. Also, additional traps should be placed in this area as it may be a hotspot. In this latter case, a careful inspection of the surrounding area is likely needed to try and determine where the spiders may be harboring.

***Insecticide Applications***—The treatments employed for brown recluse spiders are not unlike those for cockroaches. Both pests prefer to rest in cracks and voids and are secretive in their habits. Most of the insecticide applications used will involve crack and void treatments. Directed contact treatments of exposed spiders during inspections have already been discussed, but this technique may be replaced with vacuuming.

*Crack and Void Treatments*—Treatment of cracks where brown recluse might be hiding is best done using a residual insecticide dust. An inorganic dust such as silica gel works well due to its long residual life and the fact that it acts as a dessicant. Spiders, similar to all arthropods, are especially vulnerable to water loss. A dust also covers all surfaces in a crack or void increasing the chance the spider will pick up the insecticide onto its body. Dusts containing deltamethrin or beta-cyfluthrin have also been used with success in spider control programs.

Cracks can also be effectively treated with a residual aerosol insecticide. Aerosol insecticides provide quick knockdown and kill of spiders that are contacted by the application during treatment. These insecticides, however, have a shorter residual life than a dust formulation, thus making retreatments more of a possibility.

One additional step for brown recluse spiders that may be helpful involves removing all of the plates covering electric outlets and switches and treating the wall voids behind with a dust insecticide. Brown recluse spiders may move from the attic down through the walls by following electrical wires. Be sure to equip the duster with a plastic tip to avoid accidental shock.

In crawlspaces and basements, it may be necessary to apply dust insecticides into the cracks under sill plates, between floor joists that abut each other, and into the voids of foundation walls, particularly in older buildings with stone or brick foundations. A residual dust can also be applied to the sill plate areas of crawlspaces.

In attics, dust insecticides may need to be applied under insulation by lifting the insulation and applying a light amount of dust. This is time-consuming work and is only necessary when spiders are found living under the insulation. Application of dust on top of the insulation will not provide satisfactory results. It is especially important to focus on areas where light fixtures and vents are installed in the attic floor as spiders can enter the living spaces below from around such fixtures.

Some attics will have flooring placed on top of the attic joists to allow for attic storage. This can create a difficult treatment problem because insulation is usually present under the boards. This area may be treated by drilling through the floor between each attic joist. The dust can then be applied using a large hand duster or cordless electric duster. It may be necessary in some cases to drill two holes side-by-side. A wire or other such object can be used to push the insulation down away from the floorboard while applying dust through the other hole allowing the dust to penetrate further into the void.

*Spot Treatments*—The application of a liquid residual formulation (WP, SC, or CS) to areas where brown recluse spiders might crawl can be helpful when combined with the crack and void treatments. Spot treatments alone will often provide poor results.

Spot treatments for brown recluse spider control should be directed to floor/wall junctures where these spiders are likely to crawl. These treatments are not needed along open areas, rather they should be applied along walls behind items stored or placed against the walls. Spiders are more likely to remain in these protected areas and be in contact with the treated surface longer than they would be in open, exposed areas of walls.

In crawlspaces and basements, spot treatments can be applied to the sill plate/box header area at the top of foundation walls.

*Directed Space (Contact) Treatments*—ULV space treatments are likely the least effective technique for brown recluse spiders, yet may be the one of the most frequently used. A space treatment is a volumetric treatment designed to fill an enclosed space with a nonresidual insecticide. This technique is best used for controlling flying insects, not a pest that hides in protected cracks and voids.

*Exterior Treatments*—The most important control measure for outside areas is the removal of as many potential harborages as possible as discussed earlier in this section. This usually results in a reduction of spiders near the building.

Any exterior crack in the exterior walls of the building should be treated with a residual dust insecticide. The customer should then be advised to seal exterior cracks to prevent spiders from entering in the future.

A perimeter treatment may be applied to the foundation and ground away from the building, but this treatment typically has limited effectiveness for recluse spiders

***Conclusion***—Controlling and hopefully eliminating brown recluse spiders can be a time-consuming effort but is generally necessary because of the potentially dangerous nature of this spider. Any treatments are likely to make the spiders more active which may bring them into contact with persons living or working in the building. Customers should be advised of this fact and also be advised concerning practices that can help prevent direct contact with spiders. The generous use of monitoring traps will help catch and remove many of these spiders as they move about in response to treatment.

# ORB-WEAVING SPIDERS

## *Families Araneidae, Tetragnathidae* and *Uloboridae*

**Key Identifying Characters.** The orb-weavers are readily identifiable from their web which is the classic image of what people think of when you ask them to conjure up something that constitutes a spider web. It is the two-dimensional, orb-shaped web, with a continuous spiral of silk emanating from the central portion of the web with many radial spokes of supporting threads similar to the spokes on a bicycle wheel. Although a few species have evolved away from the orb web, most of these three families of spiders make an orb web.

Orb-weaver spider, *Araneus* sp.

The araneid orb-weaving spiders are most common, and the plane of their web is vertical or mostly vertical (Figure 1); these spiders are known simply as the orb-weavers.

The tetragnathid spiders are less common and used to be considered araneid spiders. They make similar orb webs although the position can be horizontal or vertical. Many of these spiders have very long chelicerae so they are known as the long-jawed orb-weavers.

The uloborid spiders are less common but can be plentiful around homes. The plane of their orb web is horizontal or slightly askew from horizontal. These spiders are known as the hackled orb-weavers because they use the fluffy cribellate silk similar to the large crevice weavers (Family Filistatidae).

Although to the casual observer, the variety of orb webs may look very similar, many of the species of these three families make characteristic webs

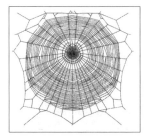

Figure 1: Orb-weavers spin distinctive flat, round spider webs.

such that one can identify the species just from the aspects of the web such as its size, whether it leaves a pie slice of the web without silk, if there is a central hub, if there is a gap of silk between the hub and catching threads, if the weaver decorates its web with dead leaves and the corpses of its prey, etc. In addition, spiders of the genus *Argiope* are well known for putting conspicuous thick swaths of silk near the center of their web that may be in the shape of a large "X". Although these are called garden spiders, they are also known by the name, "writing" spiders.

Figure 2: Orb-weavers have large round abdomens and many spines all over their legs.

Anatomically, the araneid spiders generally have large, bulbous abdomens that look extremely cumbersome, very short third legs, and often with many thick spines all over their legs (Figure 2).

Their eye pattern is also rather diagnostic. Whereas most spiders have their two eye rows with eyes about equally spaced, araneid spiders have the four middle eyes of each row close together in somewhat of a square and then, the lateral eyes are positioned quite a distance away (Figure 3).

The most commonly encountered araneid orb-weavers are those of the genera *Araneus* and *Neoscona* which have many species and are distributed throughout North America. They have rounded abdomens, sometimes with two pointed prominences near the anterior edge, and, hence, get the name "cat-head spider." Both *Araneus* and *Neoscona* usually have a diagnostic pattern of a black patch on the middle of the belly

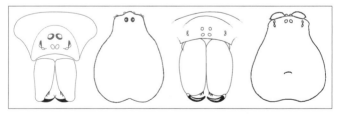

Figure 3: The lateral eyes of an orb-weaver are widely spaced from the median eyes. The line of the eye row of a garden spider (left) curves differently than most orb-weavers (right).

Figure 4: Large orb-weavers of the genera *Araneus* and *Neoscona* often have a characteristic marking of white brackets bordering a black spot on the belly.

that is surrounded on the lateral sides by two white or tan commas or sometimes four dots (Figure 4). Some of the common orb-weavers are the cross spider (*Araneus diadematus*), the shamrock spider (*A. trifolium*), and the barn orb-weaver (*A. cavaticus*); this last spider was the species that was Charlotte of *Charlotte's Web*. An infestation of *N. oaxacensis* occurred inside a warehouse in northern Mississippi that required extensive spider control efforts.

Other very common and conspicuous orb-weavers are the garden spiders. The yellow and black garden spider (*Argiope aurantia*) is very well known to homeowners, making a very conspicuous web with the X in the middle, having spectacular brilliant yellow body coloration contrasting with the black (Figure 5). Although many homeowners appreciate their beauty, others still find them too creepy and want them destroyed.

The spiny-backed orb-weaver, *Gasteracantha cancriformes*, is a colorful species found across the Gulf states, particularly in Florida where it is common around homes (Figure 6, page 186) (also see color image section in the book). The white abdomen is adorned with black spots and red spikes or spines, and unlike most spiders, it is quite hard. In some areas, the color varies with individuals having black spikes or the body being more entirely black. Their webs are found situated among the branches of trees and shrubs and on building exteriors.

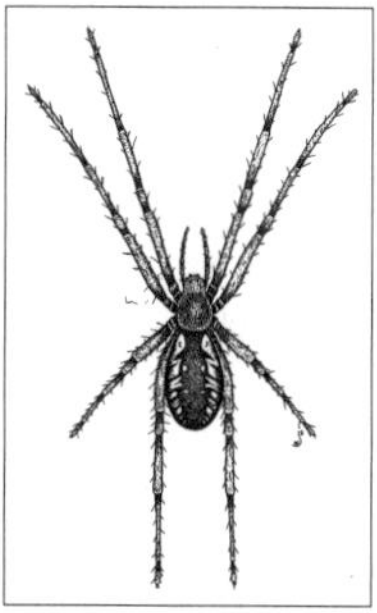

Araneid orb-weaving spiders range from 1.5 mm to about 30 mm in body length when mature, with 31 genera and 161 species in North America.

Figure 5: Garden spider.

The long-jawed orb-weavers are typically long and thin, usually making a horizontal web. They have massive chelicerae which extend outward at the base, exceedingly long fangs which are used in their mating rituals and their eyes are more evenly spaced (Figure 7, page 186). Most of the tetragnathid orb-weavers are

Figure 6: Spiny-backed orb-weaver spider.

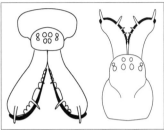

Figure 7: The long-jawwed orb-weavers have massive, curved chelicerae and fangs, however, they are used only during courtship and cannot readily bite due to the cumbersome length.

medium sized spiders, but in the southeastern United States, the golden silk spider (*Nephila clavipes*) is gigantic with females attaining a body length of 50 mm. Its silk is actually a yellowish color rather than white or transparent, and the section of spaces between the silk strands are often rectangular instead of radial slices that one would find in a typical orb web. The female also has brushes on her legs. Tetragnathid orb-weaving spiders range from 1.2 mm to about 50 mm in body length when mature with 11 genera and 38 species in North America.

The hackled orb-weaving spiders are mostly identifiable by their body form, which is very angular (Figure 8). Many species have long hairs on the first leg in the form of a brush and are in the genus *Uloborus*. Hackled orb-weaver eyes are widely and weirdly spaced (Figure 9, page 187). These hackled orb-weaving spiders range from 4 mm to about 8 mm in body length when mature with seven genera and 16 species in North America.

Figure 8: The hackled orb-weaver has an angular body with long front legs.

**Biology in General.** If you remember the children's story of *Charlotte's Web* by E. B. White, at the end, Charlotte dies and her babies hatch out the next year to start life anew. This is the actual seasonal pattern for orb-weavers, very similar to an annual plant, which grows from seed to flowering plant then dies all in one year. They start out in the spring as small spiderlings, and, within a few months, they grow tremendously, molt many times, become adults, mate, reproduce, and die with the onset of winter. This is rather amazing considering how large some of these spiders can get.

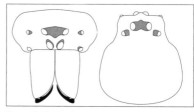

Figure 9: Hackled orb-weavers eyes are set far apart and are oriented in several directions.

Also amazing is their ability to spin a web in just a few minutes, quickly moving over the surface of the sticky silk, applying new strands. The web design, of course, allows the spiders to sweep a larger portion of the environment for food. Spider silk is renowned for its elasticity and strength. When a flying insect hits a web, the energy of flight is absorbed by the microstructure of the silk preventing it from breaking. Then the silk's elasticity allows the strand to return to its previous form so that it retains its original shape. It can also support a tremendous amount of weight considering its own mass. Because of these aspects of spider silk, scientists are studying these fibers in the hopes of developing better parachutes and bullet proof vests.

Many of the araneid orb-weavers construct a web every day. At the end of their predatory period (which can be either day or night), they take down the old web by cutting the silk strands and eating the silk to recycle the valuable protein. To reduce the web to a consumable mass, they either cut the silk at the bottom at which point the web rolls up like a window shade or they bulldoze the silk to the middle of the web into a pile.

The hackled orb-weavers are very interesting among spiders as they are one of the few families of spiders that lack venom glands. To subdue prey, they just wrap it and then consume it alive.

**Biology in Regard to Pest Control.** Orb-weavers can be extremely common around homes especially in the late summer and early autumn as the females are large, conspicuous, and their webs are strung between two supports like trees. This is the time when people, to their dismay, run into the web and get completely covered with spider silk. House eaves and porches are favorite sites for web construction.

Because many of these spiders are nocturnal, they will be hiding

Figure 10: A massive infestation of millions of orb-weavers created a particularly interesting pest control issue in a Maryland water treatment plant.

in cracks and crevices in the house eaves or other crevices so one must look for the spiders carefully. However, with day-active spiders, because they re-ingest the silk, spraying the web may actually help reduce spider populations. However, this means leaving the spider web where the homeowner can see it, which may be undesirable for all involved.

One useful biology fact about orb-weavers is because they are annual spiders, if you are able to eliminate them from a structure, there should not be a reinfestation until the following season because they do not keep reinfesting with new spiderlings as would be the continual baby production of something such as a black widow.

In an amazing story, a massive infestation of an araneid (*Larinioides sclopeteria*) and a tetragnathid (*Tetragnatha guatemalensis*) in a water treatment plant in Maryland covered 94,000 ft$^2$ and a volume of 147,000 ft$^3$ with estimates that this was caused by 100 million spiders (Figure 10). These spiders were also responsible for a massive natural infestation in a park in Texas in 2008.

**Toxicity.** The larger orb-weavers (typically araneid spiders) are big enough that if they bit, it would be painful from fang puncture. Although they have long chelicerae and massive fangs, the long-jawwed orb-weavers have no muscular strength to puncture human skin. Neither of these groups have venom that is more than mildly toxic to humans. As mentioned above, the hackled orb-weavers lack venom so they pose no threat to humans.

**Inspection Tips.** Orb-weaver spiders are generally easy to find because their characteristic flat, orb webs are usually located in open areas in and outside of buildings. These webs are common in corners, under eaves, and hanging between branches on trees. Inspections should be geared first toward identifying the presence of webs, then determining if a spider is present in the web. Many orb-weavers build new webs every evening and take the web down in the morning. The spider then hides in a crack or vegetation during the day which makes them difficult to detect during daylight.

**Key Management Strategies.** Significant infestations of orb-weaver spiders are rare inside structures, and invasions are generally limited to a few wayward spiders in windows or in corners near doors. Exceptions do occur, however, where orb-weavers are found inside warehouses or other large structures in greater numbers thus warranting significant control efforts. It is more likely to see these spiders in garages, sheds, and warehouses than inside the living areas of homes. Cellar spiders and comb-footed spiders are usually the spiders seen constructing webs inside buildings.

Figure 11: Orb-weavers can create very large webs between vertical supports such as trees, lamp posts, and sign posts.

*Sanitation*—Like most spiders, orb-weavers are present due to the presence of insects that serve as their food. An abundance of spiders indicates an abundance of insects so any steps taken to reduce the populations of insects are obviously beneficial to reducing the numbers of spiders. One of the primary ways to reduce insects is to reduce their potential harborages by removing piles of debris and cutting back heavy vegetation away from the structure.

Control of orb-weaver spiders, like the comb-footed spiders, requires regular removal of the webs, both old and new. Web removal allows the detection of new activity by the presence of new webs. Webs are most easily removed by the use of a device called the cobweb duster (Webster™) or by vacuuming. Vacuuming is preferable with larger numbers of spiders as it removes the spiders and any egg sacs along with the webs.

*Exterior Lighting*—One of the more common places to find orb-weaver spiders is outside near exterior lights (Figure 11). Metal halide lights on the outside of structures attract insects which, in turn, attract spiders, including orb-weaving spiders. Changing to sodium vapor lights on commercial buildings attracts fewer insects. Switching to the yellow "bug" light bulbs in outdoor light fixtures on homes will result in a reduction of flying insects.

***Exclusion***—Most orb-weaver spiders that enter buildings do so by crawling through cracks around windows and doors. Sealing exterior cracks in these areas as well as installing tight-fitting weatherstripping is helpful in dealing with infestations of these spiders. Although standard window screens exclude most pests, $2^{nd}$ instar spiderlings can enter through screening that is 20 mesh or larger. Keeping windows and doors closed, therefore, may be best when larger screening mesh is used on screens.

***Insecticide Applications***—Normally, the nonchemical control measures described above are sufficient for controlling orb-weaving spiders. In fact, it is likely that identification of a single specimen that has been caught by someone will be the only duty asked of a pest management professional. If numerous orb-weaver spiders are found in or around a structure, insecticide applications might be necessary.

*Directed Contact Treatment*—When found in its web, an orb-weaver spider is best removed by vacuuming when its web is removed. If a vacuum device is not available, direct application of a nonresidual aerosol insecticide to the spider hanging in its web will kill a single spider.

*Spot Treatments*—The application of a residual liquid insecticide to corners where these spiders are likely to build their webs can be useful in preventing some new spiders from becoming established. A WP, SC, or CS formulation should work well for these treatments.

# CELLAR SPIDERS

## *Family Pholcidae*

**Key Identifying Characters.** Most of the cellar spiders encountered by homeowners have very long, thin, spindly legs such that they are also referred to as daddylonglegs spiders (and also incorrectly as daddylonglegs). (Daddylonglegs are actually another Order of arachnids called Opiliones. See the chapter on *Relatives of Spiders* for more information on these creatures.)

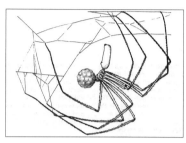

*Pholcus phalangioides*, in web, holding her ball of eggs.

Most species of cellar spiders have eight eyes with a cluster of three eyes on either side and two smaller eyes in the middle (Figure 1). One tiny pale species (*Spermophora senoculata*) has six eyes, in two clusters of three. The smaller species typically have round or globular abdomens; the larger species have elongate abdomens and one species (*Crossopriza lyoni*) originally from Southeast Asia, has a truncated abdomen where it looks like someone chopped off the hind portion. In the west, the marbled cellar spider (*Holocnemus pluchei*) is incredibly common (Figure 2, page 192); it has a dark brown sternum, which continues as a dark brown longitudinal stripe on the ventral portion of the abdomen. Male and females look similar in coloration within a species. They range from 2 mm to about 10 mm in body length when mature with 12 genera and 34 species in North America.

**Biology in General.** As the name implies, cellar spiders are often found in basements and crawlspaces of buildings where they make irregular cobwebs in window wells, next to book-

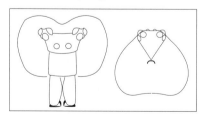

Figure 1: The cellar spider typically has eyes on turrets.

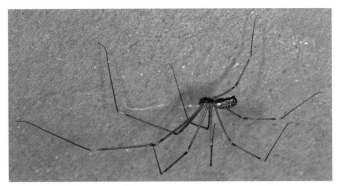

Figure 2: The marbled cellar spider is an extremely common urban pest species in the southwestern United States.

shelves in corners, and underneath tables or other horizontal surfaces. In nature, they are typically found under rock outcroppings, in caves, and underground objects. They use their long flimsy legs to safely wrap prey from a greater distance than do other spiders. Also, if disturbed as if by a predator, many of the cellar spiders will whirl around in their webs such that they are just a blur of legs and small body. Of course, a moving target is much more difficult for a predator to attack. They also will move with lightning quickness to the back of the web if they sense the wingbeat vibrations of a predatory wasp that is searching for prey.

These spiders do not recycle their silk like some other web-spinners so their flimsy webs build up, attract detritus and can be incredibly messy. They are perfect spiders for naturally decorating a house for Halloween (Figure 3).

Figure 3: This is the build-up of cellar spider webs in an abandoned outbuilding.

**Biology in Regard to Pest Control.** Cellar spiders are commonly encountered by pest control professionals and may be one of the biggest reasons for requests for pest control by homeowners. They can become incredibly plentiful in homes and businesses. Some species may lay down such thick layers of silk over a long time that they will reduce airflow into buildings or machinery, resulting in overheating of equipment.

The most common species throughout North America is a European immigrant (*Pholcus phalangioides*) which is uniformly gray all over. In the southeastern United States to Texas, the species with the truncated abdomen (*C. lyoni*) is a common pest. In the western United States, the marbled cellar spider (*H. pluchei*) is incredibly common under the eaves of house as well as indoors, making webs in the corners of rooms.

Figure 4: Cellar spiders carry their egg sac in their fangs.

One bit of biology that is useful for the pest professional is that cellar spiders carry their eggs in their fangs (Figure 4) unlike many other spiders which hide their eggs in cracks and crevices or lay eggs in a retreat so the eggs are under a protective layer of silk. In contrast, the cellar spider eggs are very loosely held together with a few flimsy strands of silk. Therefore, in spraying for cellar spiders, a pest professional is also able to kill the eggs and young spiderlings if they are not yet dispersed from the mother.

**Toxicity.** There is a very widespread and well-known myth throughout the English-speaking world that these spiders are some of the most toxic spiders in the world but their fangs are too short to bite through skin. This myth has no scientific basis. The spiders do indeed have short fangs (called "uncate") but, likewise, recluse spiders have uncate fangs and they can bite. Attempts to find the origin of this myth led to nothing substantial. However, the bottom line is that cellar spiders are not toxic, thus providing no basis for this toxicity myth.

**Inspection Tips.** Cellar spiders are generally easy to find. Even though the webs are located in dark places, they are not hidden from view, being constructed mostly in corners. Most of the time, these spiders confine themselves to a building's substructure or in the garage—they seldom venture into the living areas of a structure. If they do, they keep to dark corners behind doors and in closets. They have been found living behind the drawers and up under the sinks of kitchen cabinets or inside the bathtrap void under a bathtub. Older homes with excess moisture problems seem to experience the largest populations, probably due to the insects attracted by such conditions. It has been reported that the females of *Pholcus* do not survive well inside rooms with central heating and that her eggs do not develop and hatch properly in such rooms.

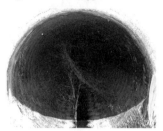

Figure 5: Cellar spiders can be extremely dense in suitable habitats such as this drain pipe.

The profuse web building by cellar spiders is usually responsible for many of the crawlspaces which seem to be "filled" inside with webs. Crawlspaces such as this are generally those with excess moisture conditions that support larger insect populations and the spiders that feed upon them (Figure 5).

In commercial buildings, cellar spiders are commonly found in corners and up in the roof beams in warehouse and other storage areas. The worst infestations seem to occur in buildings located in more rural areas. Warehouse areas become havens for spiders because workers often leave doors open that allow innumerable flying insects to enter. More insects create a greater food source for spiders with the result being more spiders.

**Key Management Strategies.** Cellar spiders are not difficult to control. If the population can be reduced first by sanitation and then limited insecticide treatments, sightings of these spiders in the living area of a building will be rare. The key to long-term control is to correct environmental conditions that attract their food—insects—and to seal exterior cracks to exclude new spiders from entering.

*Contributing Conditions*—The most important condition to address in reducing the threat of cellar spider infestations is the reduction of populations of insects serving as the spiders' food source. This can be accomplished in several ways.

*Harborage Removal*—All items lying on the ground, such as piles of lumber and other debris, stones, and boards should be removed from the property. Heavy vegetation such as ivy and other ground covers, as well as branches of trees and shrubs should be cut away from the building.

Firewood piles should be moved as far from a home as possible and stored off the ground and covered with plastic to keep it dry.

*Exclusion*—Any crack in the exterior of a building through which insects or spiders could enter should be sealed. All vents, windows, and doors should have no cracks around them and should be equipped with screens. It is also important to seal cracks around pipes, wires, and cables leading up into the structure from the crawlspace or basement.

To reduce the number of flying insects entering warehouses, overhead screen doors or plastic strip curtain doors may need to be installed in doorways that remain open for ventilation purposes.

*Exterior Lighting*—Metal halide lighting on commercial buildings can attract large numbers of flying insects which could enter the building, particularly in warehouse areas where cellar spiders are likely to be found. Changing lighting to sodium vapor lighting attracts many less insects.

*Ventilation*—The largest infestations of cellar spiders seem to be in those crawlspaces, cellars, or basements that have poor ventilation or are otherwise damp. Improving crawlspace ventilation by installing foundation vents and vapor barriers makes crawlspaces less attractive for insects. Installing a dehumidifier and/or a sump pump in a damp basement or cellar reduces the moisture that supports insect populations.

**Sanitation**—When inspecting for cellar spiders, it is helpful to carry a vacuum device, if available, with you to remove the spider webbing and live spiders as they are found. This method would be difficult to use in a crawlspace, however, so another method might be necessary. A product called Webster® or Cobweb Duster® (see *Control Techniques for Spiders* chapter) easily removes spider webs and can be used anywhere in a structure including basements and crawlspaces. When encountering a web with a live spider, however, the spider should first be treated with a directed contact treatment using a nonresidual aerosol insecticide. Exempt insecticides such as those containing essential plant oils (e.g., thyme, clove, eugenol) are also effective when applied directly to spiders. The webbing can then be removed with the cobweb duster or a similar brush.

Web removal is an extremely important step in controlling cellar spiders in order to more easily detect new spider activity during future inspections or services. It should be done on a regular basis for long-term relief from these spiders.

***Insecticide Applications***—Once cellar spider webs and any live spiders have been removed, the corners where spiders are likely to appear again may be treated with a residual insecticide. Such applications may have limited effectiveness depending on the surface being treated, the formulation used, and the limited amount of time that cellar spiders might remain in contact with a treated surface.

*Spot Treatments*—Liquid residual insecticides applied to the corners of rooms where cellar spiders might construct webs is the best treatment option to use in living areas. Use a WP, CS, or SC insecticide formulation for best results.

In crawlspaces, basements, garages, and sheds, spot treatments can be applied to the sill plate/box header area at the top of foundation walls, between floor joists, or to corners where cellar spiders are living.

*Dusting*—In crawlspaces, it may be more effective and easier to apply a residual dust insecticide to the areas of the substructure where cellar spiders are found. Dust insecticides may be more easily picked up by spiders as they construct their webs or move about in their webs. Applications of dust in overly wet conditions, however, should be avoided as they can interfere with a dust insecticide's effectiveness (except for DeltaDust® which is moisture resistant).

*Directed (Contact) Treatments*—Nonresidual insecticides and products containing essential plant oils can be effectively used to kill cellar spiders found during inspections.

*Exterior Treatments*—An important control measure for outside areas is the removal of as many potential insect harborages as possible as discussed earlier in this section. This results in a reduction of the insects spiders prey upon.

Cracks in the exterior walls of the building should be treated with a residual dust insecticide. Weep holes in brink veneer walls may be "plugged" with a product such as Xcluder.™

Spot treatments with a WP, CS, or SC insecticide around doors and windows may be useful in detering new spiders and spiderlings from entering.

# FUNNEL WEAVING SPIDERS

## *Family Agelenidae*

**Key Identifying Characters.** Funnel weaving spiders are extremely common throughout the United States. They are most easily identified by the funnel web, which is a flat sheet of tightly-woven webbing similar to a trampoline. The flat surface leads back to a retreat of some sort, usually a hole in the mortar between two bricks or a pile of rubble (Figure 1). The retreat entrance is usually circular or oval in shape, which is where the spider can often be seen as it waits for prey to land on the web.

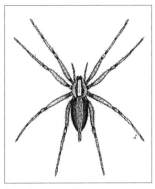

Funnel weaver spider.

Anatomically, funnel weaving spiders are somewhat generic looking in tans and browns, often with longitudinal stripes on the cephalothorax and a stripe or undulating pattern on the dorsal surface of the abdomen. The legs may have darkened rings around segments or they may be uniformly colored. They are often mistaken for wolf spiders. However, the feature that is diagnostic for most spiders in this family is the eye pattern, which is referred to as a 2-4-2 pattern. In most spiders with eight eyes, they are arranged in two relatively straight rows, or in the case of jumping and wolf spiders, the posterior eye row is curved backwards. However, in funnel weaving spiders, the two rows are on the vertical face of the spider and are so strongly curved forward that the eyes

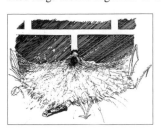

Figure 1: The funnel web is a very easy way to identify a funnel web spider. However, they are very quick to retreat into the hole and most often escape capture.

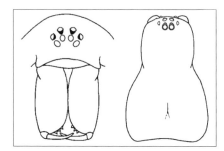

Figure 2: Most funnel web spiders have their two eye rows so strongly curved that it looks like three rows. This is referred to as a 2-4-2 eye pattern.

on the end of the front row appear to be in a row by themselves, and the last row has two eyes which are the middle eyes from the posterior row. The middle row has four eyes, which are the middle eyes from the front row and the lateral eyes from the posterior row (Figure 2).

This eye pattern is common throughout most of the funnel weavers that a pest professional will encounter with one exception. Spiders of the genus *Tegenaria*, which include the hobo spider, the giant house spider, and the barn-funnel weaver (or common house funnel weaver), have eyes appearing in two rows that are only slightly curved and which look more like the generic eye pattern of other common spiders (Figure 3).

Many of the species of funnel weavers have extremely long tapered spinnerets that extend from the posterior tip of the abdomen and can also be used to identify some members of the family. Some of these funnel weavers with long spinnerets include members of the genus *Agelenopsis* with *A. naevia* and *A. pennsylvanica* being common in the east and *A. aperta* and *A. potteri* being common in the west. *Agelenopsis* spiders often have an elongate leaf-like pattern on the dorsal surface of their abdomens. Many species of *Hololena* occur in the western United States around homes, most notably, *H. curta* in Southern California and *H. nedra* from central California up through the Pacific Northwest. *Hololena* have much shorter spinnerets and a

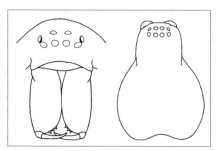

Figure 3: Spiders of the genus *Tegenaria* make funnel webs but their eyes are in the more common arrangement of two rows of four, although the rows are slightly curved.

broad longitudinal band or stripe on their dorsal abdomen (Figure 4). The *Tegenaria* spiders are also common with the small and common *T. domestica* being found throughout the United States (as well as around the world), the hobo spider, *T. agrestis* found in the northwestern quadrant of the United States, and the giant house spider, *T. duellica* (although there is some debate whether its correct name should be *T. gigantea*) restricted to the Pacific Northwest, west of the Cascade Mountains. Funnel weaving spiders range from 4 mm to about 18 mm in body length when mature with nine genera and 85 species in North America.

**Biology in General.** The funnel weavers make their webs in many places. They are also known as grass spiders because they make their funnel webs in grass as well as other vegetation. The web is constantly having silk added to it so it thickens over time. The retreat is open on both ends so if you attempt to catch the spider, it will scoot out the escape hole in the back; funnel weavers are sometimes very difficult to catch unless you can lift a board that has the entire retreat on the underside so the spider can't easily run off. When on the alert, the spider will wait in the mouth of the funnel for prey to fall on to the trampoline-like surface, whereupon it rushes out to subdue the prey and drag it back to its retreat for dinner. There are also thin silk lines running from the matting of silk to supports upward which also serves to trip up flying insects so they fall on to the silk.

Figure 4: Spiders of the genus *Hololena* are very common in the western United States.

The spiders respond to vibrations on the web. In one instance, one of the authors was playing a banjo while standing by a window, which had a funnel weaver web in the corner. As the strings were being plucked, the spider ran out from its hole and raced around the web trying to find the source of the vibrations. When the plucking stopped, the spider ran back into its hole. Resuming of playing got the spider out to dance a jig once again around the web.

**Biology in Regard to Pest Control.** Funnel weaving spiders make their webs around and inside homes. The wells of basement windows are

Figure 5: Funnel web spider populations can be very dense in suitable habitat.

favorite spots for them to set up a web site. Typically, they are low to the ground, but if the architecture of a structure is sufficient to support a web, they can also be located overhead. These spiders also can be extremely common, especially in fields where their density is highly conspicuous when the webs are covered with dew (Figure 5).

Because of historical incrimination of medical significance, the hobo spider is the funnel weaver of greatest concern although it is probably falsely accused (see *Toxicity* below and the chapter on *Health Aspects of Spiders*). The three *Tegenaria* spiders (hobo, giant house, barn funnel weaver) are all European natives that have become established in North America. The barn funnel weaver (*T. domestica*) is found throughout North America. The hobo spider (Figure 6) is known from British Columbia through Oregon and east to Montana, Wyoming, Colorado, and northern Utah and is still spreading. The giant house spider is found west of the Cascade Mountains in the Pacific Northwest and appears to be more common than the hobo spider. The hobo and giant house spiders are very similar looking; typically, one needs a microscope to examine the reproductive structures to verify the identity.

**Toxicity.** The hobo spider was incriminated in the 1980s as a spider of medical importance, however, the toxicity of its venom is being severely challenged by medical arachnologists in the 21st century. Many discrepancies exist that

Figure 6: The hobo spider's toxicity has been seriously questioned. It will require more research with a series of verified bites to determine whether it is harmless or toxic.

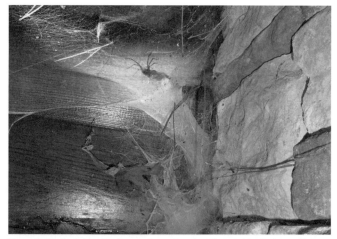

Figure 7: This large funnel weaver web was found on the outside of a home. Note the spider in the top of the web.

suggest that this spider may have been erroneously elevated to medical significance. See the chapter on *Health Aspects of Spiders* for a more detailed description. The large spiders of the genus *Agelenopsis* have caused bites with minor symptoms.

This family of spiders should NOT be confused with the Australian funnel-weaving spiders, the latter of which are incredibly toxic to humans and are not at all closely related to agelenid spiders taxonomically.

**Inspection Tips.** Funnel weaver spiders are generally easy to inspect for due to the size and accessibility of their webs. Also adding to easier inspection is the fact that these spiders generally confine themselves to the lower levels of the building. The time spent inspecting the attic and upper floors should be minimal.

Inside, the webs of the funnel web spider will be confined to those areas where sufficient items exist to support the funnel portion of its web. Look for holes or cracks in walls from which extends a flat web. Look for webs extending out from stacks of lumber or other items and from between stored boxes. The webs of the grass spider, *Agelenopsis*, and the domestic house spider also can be found in corners of rooms, closets, and shelving.

Monitoring traps can be useful during the summer and fall to capture males in search of mates and wandering immature spiders. The traps also serve as a minimal control method removing individual spiders as they wander inside.

Outside, the funnel weavers will be found by looking for their webs located on buildings (Figure 7, page 201) in tall grass, ground cover, and among debris, rocks, and other items. Webs can also be located along fences, in cracks in walls, under decks, and in sheds and other outbuildings.

**Key Management Strategies.** Control of funnel weaver spiders found inside are likely necessary as their presence inside will probably be noted due to the large webs and will be unwanted by the building's occupants.

*Contributing Conditions*—The most important condition to address in reducing the threat of funnel weaver spider infestations is the removal of potential hiding places near the foundation of the building. All items lying on the ground, such as piles of lumber and other debris, stones, and boards should be removed from the property. Firewood piles should be moved as far from a home as possible and stored off the ground and covered with plastic to keep it dry. Heavy vegetation such as ivy and other ground covers should be cut away from the foundation. Tall grass should be kept cut low. In situations where a field is next to or near the building, regular mowing of the field can be of help. These measures also reduce the potential harborages for the insects serving as the spiders' food sources.

*Exclusion*—Cracks in the exterior of a building through which spiders could enter should be sealed to prevent spiders from entering. Cracks or holes where spider webs are located are likely to be harboring spiders and should be treated as described below. Doors should have a tight-fitting weatherstrip on the bottom, especially garage doors.

*Sanitation*—Removing potential outdoor harborages for spiders and sealing exterior cracks as discussed above are both part of a good sanitation program for funnel weaving spiders. Funnel weavers also can be common within shrubs near homes and buildings. Removing shrubs is generally not feasible, and if the customer desires that spiders be controlled, then they may be treated with a spot treatment.

It is important to remove the webs of these spiders as part of the control program. This allows one to more easily determine new activity during future inspections. It is quicker to use a vacuum device for removing webs and spiders. A device like a cobweb duster (see the *Control Techniques* chapter) or a broom can also be used to remove webs. It must be noted that where webs are constructed in association with cracks or holes in walls, the spiders will likely retreat into the void behind the web and will need to be killed by treating inside the wall. The collection bag to the vacuum device should be removed outside and discarded if live spiders were also removed with webs.

It is also important to improve storage practices in garages, basements, and closets to limit the available harborages that can be used by the hobo or domestic house spiders. Store items off the floor and away from walls and reduce as much clutter as possible.

***Insecticide Applications***—Inside, treatment of live, exposed spiders can be accomplished by directed contact treatment. Spiders hiding within voids should be treated using a residual dust insecticide. Spot treatments may also be useful for controlling wandering male or immature spiders.

*Crack and Void Treatments*—Cracks inside where house spiders are living or could enter the living areas of the building from inside walls should be treated with a residual dust insecticide.

In crawlspaces and basements, it may be necessary to apply dust insecticides into the cracks under sill plates and into the voids of foundation walls where webs are located. This is particularly necessary for older buildings with stone or brick foundations.

*Spot Treatments*—The application of liquid residual insecticides to baseboard areas behind and under furniture in rooms where numerous spiders have been noted can be helpful in potentially controlling wandering males and immature spiders. Treatment of the floor/wall juncture in basements and garages may also be beneficial. Use a WP, SC, or CS formulation for best results..

*Directed (Contact) Treatments*—Nonresidual aerosol insecticides can be applied directly onto spiders found during inspections.

*Exterior Treatments*—The most important control measure for outside areas is the removal of as many potential harborages as possible as discussed earlier in this section. This automatically results in a reduction of spiders and the insects they prey upon near the building.

Cracks in the exterior walls of the building should be treated with a residual dust insecticide and then sealed by the customer to prevent spiders from entering in the future.

A perimeter treatment may be applied to the foundation and vegetation away from the building. It may be more effective to look for webs and treat individual spiders found using a directed contact treatment or with a spot treatment applied with a compressed air sprayer.

# COMB-FOOTED SPIDERS EXCLUDING WIDOWS

## *Family Theridiidae*

**Key Identifying Characters.** The members of the comb-foot spider family are diverse with many species. The black and brown widows are also included in this family, but because of their medical importance and greater concern, they are treated in a separate chapter in this book.

The comb-foot spiders are, not surprisingly, defined by a comb-foot, which is a specialized row of curved spines on the ventral surface of the fourth tarsi (Figure 1). The spines are thicker than other hairs around them, typically are arranged in a single row, and have bumps on the lower surface (serrated).

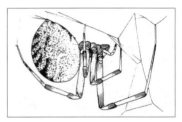

House spider.

Comb-foot spiders use these serrated spines to manipulate the silk as it is emitted from the spinnerets. The comb-foot is actually rather easy to see under a decent microscope, but it is not a useful morphological feature for a pest professional. In addition, because this is a silk spinning device, once males mature, the comb-foot becomes vestigial or disappears altogether, which can be a big source of frustration for someone trying to learn spider identification.

However, most comb-footed spiders of interest to the pest profes-

Figure 1: The specialized curved row of spines on the tarsi of the hind legs are used by comb-foot spiders to manipulate silk.

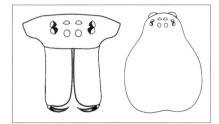

Figure 2: The eyes of a comb-foot spider are typically about the same size in two rows of four and in many species, the lateral eyes are very close to each other.

sional are similar to the black widow spiders in body form and behavior. They have globular abdomens and hang upside down in their webs. Their eyes are typically all about equal in size and are arranged in two rows of four (Figure 2).

One of the most common house spiders is, appropriately enough, called the common house spider. That is good because its scientific name is *Parasteatoda tepidariorum*. Its name was recently changed from *Achaearanea tepidariorum*, which really doesn't make it any easier to pronounce or learn how to spell correctly. This spider is drab dusky brown and tan with nebulous chevron markings on its abdomen. It is found in basements, water heater closets, under house eaves, and typically hides in leaves and detritus that it incorporates into the web. Its egg sac is tan-brown and looks somewhat like a small dirt clod. Another group of comb-foot spiders are much darker, chocolate brown in color, and can be either all dark brown or with light tan spots, triangles, diamonds, etc. These are spiders of the genus *Steatoda* which include the false black widow (*S. grossa*), which is almost as big as black widows and found on the Pacific Coast and in the eastern United States. The triangle steatoda spider (*S. triangulosa*) with several tan diamonds on its back is found throughout much of North America and *S. borealis* is common in the eastern part of the continent. Excluding the widow spiders of the genus *Latrodectus*, comb-foot spiders range from 1 mm to 10 mm (⅜ inch) in body length when mature, with 31 genera and 229 species in North America.

**Biology in General.** Many comb-foot spiders spin a tangled web of haphazardly appearing silk, but the web has a definite structure to it with a central area for the spider to rest (Figure 3, page 207). Although the silk is dry for most of the structure, some comb-foot spiders add sticky globules to stretched silk strands near the ground. If a prey item gets stuck in the adhesive silk, the spider will nip the silk near the ground at which point the stretched silk springs upward, suspending the prey in mid-air, preventing its escape.

**Biology and Pest Control.** Comb-foot spiders are common in homes, basements, outbuildings, garages, and crawlspaces. In particular,

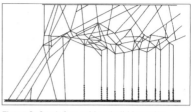

Figure 3: Comb-foot spiders build irregular, scaffold-type webs.

spiders of the genus *Steatoda* do not appear to need a lot of insect prey in order to sustain themselves; therefore, they are found under kitchen cupboards, in closets, and in the living and storage space inside a home.

**Toxicity.** Bites from comb-foot spiders other than widow spiders are not particularly nasty but some of them can still be of minor concern. False black widows are the largest of the *Steatoda* spiders, being only slightly smaller than the average black widow. False black widows survive well in houses and will inflict bites. The large females can cause pain similar to a mild black widow bite. Curiously, false black widow bites respond to black widow antivenom although this is a case of overkill remedy. One of the authors was bitten by a false black widow male while in bed; the wound was a small painless raised pimple that was hardened and slightly red for two weeks. The common house spider also can cause bites with minor symptoms.

**Inspection Tips.** Finding comb-footed spiders is generally quite easy if one takes the time to inspect carefully. House spiders will be found in open corners and in windows as well as in protected areas. Often, the only visible evidence is the old webbing left behind when the spider dies or abandons a web. These cobwebs may accumulate in significant numbers if not regularly removed, especially outside on a building's exterior or in garages, crawlspaces, basements, and storage/warehouse areas.

Inside homes and apartments, comb-foot spiders are common behind doors, between boxes, in closets and cabinets, and under shelves.

In commercial buildings, skylights, inside pallets, storage racks, and door rails are common sites to find comb-foot spiders – often they will share spaces with other spiders such as cellar spiders and sometimes, orb-weaver species.

Outside, harborage sites include under eaves, around light fixtures, under window sills, in corners of windows and doors, and behind gutter downspouts.

**Key Management Strategies.** Comb-foot spiders warrant control due to the unsightly cobwebs they produce. The control procedures for black widows will be more complete than for house spiders although the same techniques are used to control both. Other species of comb-footed spiders found inside buildings are usually most easily controlled by removal through vacuuming.

*Sanitation*—The most important item to address in reducing comb-foot spider infestations is the removal of potential hiding places outside the building. All items lying on the ground, such as piles of lumber and other debris, stones, and boards should be removed from the property. Firewood piles should be moved as far from a home as possible and stored off the ground and covered with plastic to keep it dry. Heavy vegetation such as ivy and other ground covers should be cut away from the foundation. Tall grass should be kept cut low. These measures also reduce the potential harborages for the insects serving as the spiders' food source.

It is easiest to remove existing spiders by carrying a vacuum device, if available, along during the inspection. Spiders, as well as new and old webs, can be quickly removed as they are found. Removal of webs also makes it easier to determine if new activity is present during future inspections. When vacuuming comb-foot spiders, remove the vacuum bag as soon as possible after finishing. Be careful when removing the bag and seal it in a plastic bag before disposing in an *outdoor* trash receptacle, preferably not the customer's.

It is also important to improve storage practices in garages, basements, and closets to limit the available harborages that can be used by comb-foot spiders. Store items off the floor and away from walls and reduce as much clutter as possible.

*Exclusion*—Cracks in the exterior of a building through which spiders could enter should be sealed. Doors should have a tight-fitting weatherstrip on the bottom.

*Insecticide Applications*—Spot treatments to corners can help reduce the threat of comb-foot spider infestations although this may not be totally effective.

*Directed Contact Treatments*—If a vacuum device is not available to remove spiders, treatment of live, exposed spiders can be accomplished by directed contact treatment using a nonresidual aerosol insecticide or a product containing essential plant oils.

*Dust Applications*—In crawlspaces, it may be necessary to apply residual dust insecticides into corners, onto sill plate areas, and other areas where spiders are found. This application is necessary only in those cases where numerous spiders are present in the crawlspace. Dust is effective because it clings to the webs where the spider is more likely to contact it.

*Spot Treatments*—The application of liquid residual insecticides to corners, behind and under furniture, behind stored items, and other likely web sites can be helpful in preventing new spiders from becoming established. A WP, CS, or SC formulation will generally provide the best results.

*Exterior Treatments*—Comb-foot spider infestations are best prevented by reducing potential outdoor harborages and sealing cracks and holes through which they might enter. Install sodium vapor or yellow "bug" lights to attract fewer flying insects. These steps result in a reduction of spiders and the insects they prey on near the building.

Cracks in the exterior walls of the building should be treated with a residual dust insecticide.

A perimeter treatment may be applied to the foundation and to soffits, downspouts, and similar sites where spiders construct webs. This treatment provides temporary relief and may need to be reapplied regularly in those situations where comb-foot spiders are numerous.

# WIDOW SPIDERS

## *Family Theridiidae*
## *Genus Latrodectus*

**Key Identifying Characters.** Widow spiders are comb-foot spiders, but due to their special medical importance, they are being given their own chapter. However, the anatomy for these spiders is similar to that of the other less- or non-toxic comb-foot spiders. Indeed, they have a comb-foot composed of a row of serrated curved hairs on the ventral surface of the fourth tarsi (see Figure 1 in *Comb-foot Spiders*); however, considering the diagnostic coloration of female widow spiders, it will hardly be necessary to examine the legs to confirm identification. Using this may still be necessary for the juveniles, which look very different than the females. Widow spiders have shiny, globular black abdomens with a red hourglass on the belly (Figure 1, page 212). The hourglass may not be distinct in some specimens, sometimes appearing as two spots (or in rare instances, being almost completely missing), or the dorsal side of the abdomen may also have red markings. They have no spines on any of their legs, and the third pair of legs is much shorter than the others. Their eyes are all about equal size in two slightly curved rows with the lateral eyes separated by about one eye diameter (Figure 2, page 212); in many other comb-foot spiders, the lateral eyes are much closer and sometime are touching each other (see Figure 2 in *Comb-foot Spiders*). Using this lateral eye distance has been useful in the past in trying to determine if a smashed specimen was a black widow or not.

Black widow spider.

Five species of widow spiders are found in the United States. This chapter will discuss the western and southern black widows in detail along with the brown widow, which is becoming a major pest control concern in the southern United States. The northern black widow (*Latrodectus variolus*) is found in the upper Midwest such as Michigan

Figure 1. A female black widow spider.

and Ohio but is rare and makes its webs in shrubs and trees. The red-legged widow (*L. bishopi*) is brilliantly colored with bright orange-red legs and cephalothorax and large red elliptical markings on the dorsal abdomen; this spider is only found in palmetto groves in peninsular Florida around Tampa. Neither of these latter two spiders is associated with homes, thus they will not be considered further here because they are not of interest to a pest professional. The three widow species of greatest concern are the western black widow (*L. hesperus*), the southern black widow (*L. mactans*), and the brown widow (*L. geometricus*). These three spiders are synanthropic and will generate the majority of requests for pest control services involving widow spiders.

*Western black widow*. The mature female western black widow has the well-known coloration of the red hourglass on the belly with shiny jet-black body parts everywhere else. However, it starts off life looking very different. The spiderlings, as they emerge from the yellow egg sac, have tan legs with faint dusky rings on its front legs, a tan cephalothorax with a black longitudinal stripe down the middle, a white abdomen with six to eight black dots on the sides, and a white flared shield-like shape on the belly where the hourglass develops (Figure 3, page 213). As the spider molts and grows, the rings on the legs darken and expand, the white abdomen starts to fill in with tan coloration, a white longitudinal stripe on the dorsal abdomen, white lateral stripes become evident on the flank, and the shield on the belly starts to pinch in at the middle and turn yellow (Figure 4, page 213). Additional maturation

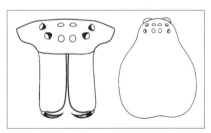

Figure 2: Widow spider eyes are all about the same size and there is a space about the diameter of one eye between the two lateral eyes.

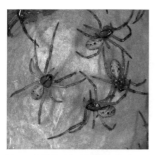

Figure 3: Baby black widows look very different from the adults.

Figure 4: As black widows mature, they typically have a pattern of light stripes.

Figure 5: A male western black widow retains the coloration of the juvenile.

results in more black pigmentation being laid down on the legs and cephalothorax, the white abdominal stripes become thinner and may develop a vivid red inner stripe surrounded by the white, and the hourglass pinches in further and turns orange. With additional molts, the white stripes turn into isolated, nebulous islands of vague white, the hourglass becomes more red, and the spider looks more like the adult female. Eventually, the spider turns completely black (although in some specimens, the female retains two white chevrons of stripes on the front surface of the abdomen) and is the stunningly vivid creature characteristic of the widows. Male spiders are much smaller in body length although legs may be nearly as long as the females. Males retain the striped coloration of the juvenile with tan or greenish-gray body coloration (Figure 5). Although on rare occasion, this base coloration may actually be black.

*Southern black widow.* The southern black widow has many parallels in development to the western black widow, but the differences will be described here. The egg sac is gray, and the spiderlings that emerge have abdomens that have white stripes on a black or dark background. As the spider grows, the light coloration gives way to darker pigment and the hourglass turns more red. Similar to the western black widow, the southern black widow is almost entirely black with a red hourglass but also has a little red dot just above the spinnerets on the dorsal surface of the abdomen (Figure 6, page 214). The male is black with white stripes.

Sometimes a western black widow retains the little red spot above its spinnerets as well as a remnant of the longitudinal stripe and is mistaken for a southern black widow.

*Brown widow spider.* The brown widow spider does not go through as drastic changes as it matures from a small spiderling to adult. The basic coloration pattern of stripes on the abdomen remains although these spiderlings have brownish cephalothoraces. As the spider ages, the typical background coloration often becomes a pale brown and resembles the parchment of tan tree bark or the outer covering of a yellowjacket wasp nest (Figure 7). In some brown widows, though, the body coloration can be extremely dark such that stripes and spots are barely visible. One key identification feature is at the top of the lateral stripes, large, squarish black dots are present (Figure 8). In the immature black widows, these dots are elliptical and small. These brown widows are frequently mistaken for immature western black widows.

*Egg sacs.* The egg sacs of these three species are actually much easier to identify than the spiders. The western black widow makes a round or teardrop-shaped yellow sac with distinct (not fuzzy) margins, and the silk is extremely difficult to rip (Figure 9, page 215). The southern black widow makes a round, gray egg sac. The brown widow makes a yellow egg sac covered with silk spicules (spikes) (Figure 9, page 215) so it looks like an enlarged

Figure 6: A female southern black widow feeding on a scorpion.

Figure 7: A female brown widow spider. There is much variation in the abdominal coloration. This female is a very pale specimen.

Figure 8: The large dark rectangular blocks on top of the lateral stripes are valuable features for identifying a brown widow.

Figure 9: Egg sacs of the western black widow (left) and brown widow (right).

pollen grain or a floating harbor mine from World War II.

The scientific names for the widow spiders have changed over the years. In the older literature, one can run across the name *Latrodectus mactans* for the widow spiders throughout North America into Central America where the color differences were considered to be just variation within one widely-distributed species. It has since been split into the several species we know today, but because people of many backgrounds write about widow spiders, they will often grab an old, outdated name, which is incorrectly cited. Also, for some unknown reason, people want to put an extra "c" in the name and come up with "*Lactrodectus*" as if this was a "milk spider" or something similar. This mistake has shown up in authoritative textbooks and medical papers. The name is *Latrodectus,* and it means "secret biter" which aptly describes how this spider envenomates people. Widow spiders range from 2.5 to 6.5 mm in body length for the males and 8 to 15 mm for the females. One genus and five species occur in North America.

**Distribution.** The southern black widow is distributed mostly from the southern states south of Maryland, west to Kansas and everything southward. The western black widow can be found from the Plains states west to the Pacific Coast, even at high elevation, which gets annual snow cover as in Denver but also thrives in Phoenix and the hot southwestern deserts. The brown widow was restricted to peninsula Florida for decades, but in the first decade of the 21$^{st}$ century, it has quickly spread and now is firmly established from Texas to South Carolina. It is now a common spider in urban Southern California and still expanding its range.

**Biology in General.** Widow spiders are mostly nocturnal although the small, lighter colored immatures can be found in the sunlight. They build a haphazard looking web, usually close to the ground although if there are enough supports high up (e.g., the exposed beams of a barn or around skylights in a warehouse), they have no problem making webs over one's head as well. They make their webs near the ground so they can catch ground-dwelling insects as well as those that fly.

The name "widow" was given to the spider because of the thought that they always eat the male after mating. This does happen in Australian species where the male does an obligatory backflip while mating and offers his abdomen to the female for nourishment, but in the North American species, the male prefers to run off if he can so he can find more females with which to mate. The "widow" name likely got tacked on to the North America spiders by sensationalist news media and, also if a male was thrown into a jar to view the mating and thus not able to escape, the female ate him.

**Biology in Regard to Pest Control.** For the pest professional, identification is important if more than one species of widow spider is found in your area. Egg sacs are very useful for species identification although one needs to be aware if the false black widow may also occur locally. One interesting aspect of black widows is that their silk makes an audible sound when ripped; the louder the sound, the larger the spider so one can even give an estimate of the size of the spider.

**Toxicity.** The toxic reputation is well-known and well-deserved. It was responsible for deaths before medical care improved. The black widow spiders can deliver a potent bite, which leaves their victim in quite a bit of pain. The brown widow bite seems to be less toxic where the most common symptoms are pain at the bite site and some redness. The bites of these spiders are covered more extensively in the chapter on *Health Aspects of Spiders*.

**Inspection Tips.** Finding widow spiders is generally quite easy if one takes the time to inspect carefully. Widows typically prefer protected sites for web location. Outside, they will be found within and under piles of debris, pipes (Figure 10, page 217), lumber, stones or firewood, behind heavy vegetation (especially against buildings), at the base of low-growing shrubs, inside water meter and irrigation boxes, under and behind grills and lawn furniture, and similar locations. Many bites occur when a female widow crawls into a shoe or boot left on the porch or garage. Spiders will also be found amongst items stored within sheds and underneath a shed, deck, or doghouse.

Service professionals checking rodent bait stations in the south and west need to be careful opening the stations as widows highly prefer the protected, dark space the boxes provide.

Inside homes and apartments, widow spiders may be found between boxes, inside closets and cabinets, and under and furniture, and underneath shelving. They will be found in crawlspaces around debris, piers and pipes and commonly build webs along the sill plate and box headers. Care needs to be taken when entering a crawlspace as widows and other spiders are often present near the openings. Widows can be common in garages in regions such as southern California, Phoenix, Tucson, and along the Gulf Coast.

In commercial buildings, skylights, inside pallets, storage racks, and door rails are common sites to find widows – often they will share the same general area with other spiders, such as cellar spiders and sometimes, orb-weaver species.

Figure 10: Black widows prefer to make their web from a retreat low to the ground. Here is a widow web coming out of a pipe. Much of the vegetation is suspended in the web.

Wearing gloves when conducting inspections for widows to avoid accidental bites is recommended. Turn over items to look for webs and spiders. It is beneficial to be prepared to treat or vacuum spiders as they are uncovered.

**Key Management Strategies.** Widow spiders warrant control because they pose a serious health threat. People become very concerned when even one widow spider is found in or near their home or place of work. The control procedures for widows are more involved than for other comb-foot spiders although the same techniques are used to control both. Regular inspections and/or treatments may be necessary in many situations where widow spiders are abundant.

*Sanitation*—The most important item to address in reducing widow infestations is the removal of potential hiding places outside the building.

All items lying on the ground, such as piles of lumber and other debris, stones, and boards, should be removed from the property. Firewood piles should be moved as far from a home as possible, stored off the ground, and covered with plastic to keep it dry. Heavy vegetation such as ivy and other ground covers should be cut away from the foundation. Tall grass should be kept cut low. These measures also reduce the potential harborages for the insects serving as the spiders' food source.

It is easiest to remove existing spiders by carrying a vacuum device, if available, along during the inspection. Spiders, as well as new and old webs, can be quickly removed as they are found. Removal of webs also makes it easier to determine if new activity is present during future inspections. When vacuuming widows, remove the vacuum bag as soon as possible after finishing. Be careful when removing the bag and seal it in a plastic bag before disposing in an *outdoor* trash receptacle, preferably not the customer's.

It is also important to improve storage practices in garages, basements, and closets to limit the available harborages that can be used by widow spiders. Store items off the floor and away from walls and reduce as much clutter as possible.

*Exclusion*—Cracks in the exterior of a building through which spiders could enter should be sealed. Doors should have a tight-fitting weatherstrip on the bottom. Windows and foundation vents should be equipped with tight-fitting screens. The mesh size of the screen is important, however, as $2^{nd}$ instar black widow spiderlings are able to squeeze through mesh screen of 0.83 mm openings (20 mesh) or larger. Screening of at 0.59 mm (30 mesh) or smaller will effectively screen out widow spiderlings (Figure 11). However, one warning to keep in mind, that this mesh size is extremely small and may be completely unusable in real world situations. If there is significant airflow through the screen for ventilation or cooling, the screens may have to be cleaned frequently.

Figure 11: This mesh (as compared to a penny) is capable of excluding baby black widows, however, it is so small that it is impractical for installation in many real-world situations.

*Lighting*—Bright, white lighting on a building's exterior is highly attractive to night-flying insects thus serving up a nightly smorgasbord for widows and other spiders. Simply switching exterior fixtures to yellow or sodium vapor lamps is highly beneficial in overall spider management.

***Insecticide Applications***—Insecticide applications can be helpful in widow spider control programs to help reduce or prevent new infestations.

*Directed Contact Treatments*—If a vacuum device is not available to remove spiders, treatment of live, exposed spiders can be accomplished by directed contact treatment using a nonresidual aerosol insecticide.

*Dust Applications*—In crawlspaces, it may be necessary to apply residual dust insecticides into corners, onto sill plate areas, and other areas where spiders are found. This application is generally necessary in those cases where numerous widow spiders are present in the crawlspace. Dust clings to the webs where the spider is more likely to contact it.

*Spot Treatments*—The application of liquid residual insecticides to corners, behind and under furniture, behind stored items, and other likely web sites may be helpful in preventing new spiders from becoming established. A WP, SC, or CS formulation will generally provide the best results.

*Exterior Treatments*—Widow spider infestations are best prevented by reducing potential outdoor harborages, and sealing cracks and holes through which they might enter. These steps result in a reduction of spiders and the insects they prey on near the building.

Any exterior crack in the exterior walls of the building should be treated with a residual dust insecticide and then sealed to prevent spiders from entering in the future.

A perimeter treatment may be applied to the foundation and the ground up to 10 feet away from the building. This treatment generally provides only temporary relief and may need to be reapplied regularly in those situations where widow spiders are numerous.

# MESHWEB AND FLATMESH WEAVING SPIDERS

## *Families Dictynidae* and *Oecobiidae*

**Key Identifying Characters.** These two spider families are considered together because they are tiny (1 to 4 mm long), found in homes, and probably would be indistinquishable to both homeowners and pest professionals alike. These mesh weavers are in a special grouping called cribellate spiders (similar to the crevice weavers, *Kukulcania*). They spin a dry silk from a modified spinneret field (anterior to the normal spinnerets) that has hundreds of little spigots instead of just a few large spigots which produce dragline silk as seen in the other spiders. Because of the field of spigots, under the microscope, the silk looks like skeins of yarn from a craft store. The silk is dry but acts like Velcro® in that it entangles the legs of prey instead of using adhesive like that of other spiders.

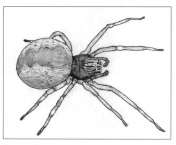
Meshweb weaver spider.

From an anatomical perspective, the feature that one would look for is a specialized spinning plate near the posterior end of the body called a cribellum. Along with this, every spider with a cribellum has a specialized row or two rows of spines on the fourth metatarsus called a calamistrum. In classic arachnological taxonomy, one would look for these features with a microscope, but because these spiders are so small, this is out of the realm of the pest professional. The other problem with attempting to find a cribellum or a calamistrum is that because these are related to silk production, the males lose these structures or they become vestigial upon maturity because at that point, the male no longer maintains a web and just searches for females.

Therefore, the best way to identify these spiders is by body

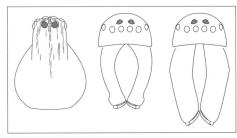

Figure 1: Meshweb spider carapace and face of male (middle) and female (right).

form. The meshweb weavers often have conspicuous white hairs pointing forward and covering most of the cephalic region near the eyes; also the posterior median eyes are usually reflective and look silver in color (Figures 1 and 2). The flatmesh weavers have eight eyes clustered tightly near the front of the carapace, some eyes larger than others (Figure 2). But they also have short, striped legs that they often hold such that the femur of each leg is perpendicular to the ground (Figure 3, page 223). The cephalothorax is also very rounded and has spots.

The meshweb weavers range from 1 mm to about 4 mm in body length when mature. In North America, 20 genera and about 290 species occur. The most common of these spiders that would be found around homes are of the genus *Dictyna*. The flatmesh spiders range from 1 mm to about 3 mm in body length when mature with two genera and eight species found in North America. The most common flatmesh weavers belong to the genus *Oecobius* with *O. navus* (formerly known as *O. annulipes*) being a non-native and ubiquitous.

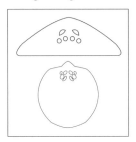

Figure 2: The eyes of the flatmesh weaver are clumped together with one pair of eyes being flat and often not easily visible. It also has a roundish cephalothorax.

**Biology in General.** The flatmesh weavers run quickly and randomly as if they have no idea where they are going. However, if they run into an ant, they immediately switch from frantic, unorganized movement to running tight circles around the ant, laying down silk and tying the ant to the ground. Sometimes, they reverse direction and make circles the other way to ensure the ant doesn't escape.

**Biology in Regard to Pest Control.** These spiders spin their fluffy silk webs flat against houses and in window sills (Figure 4). They are very commonly embedded in stucco where many nooks and crannies exist to create suitable web sites. They also will be in great abundance around outdoor lighting. Their webs can be very unattractive as they gather dirt and other airborne detritus.

Figure 3: The flatmesh weaver has striped legs which it often holds at an awkward angle.

**Toxicity.** These spiders are far too small to cause any medical damage whatsoever.

**Inspection Tips.** Meshweb and flatweb spiders prefer the outdoors and only wander inside on occasion where they are often found on window sills. Homeowners are more likely to be concerned about the unsightly webbing on the home's exterior.

Inspections are fairly straightforward as the webbing gives away the location of both active and old harborages. Check the building exterior and inside the garage or even inside sheds or other outbuildings for webbing and spiders.

**Key Management Strategies.** These spiders generally warrant control due to the unsightly webs they can produce which, although small, can be extremely numerous. Most of these efforts will be directed at the walls of the building exterior particularly around exterior light fixtures and window sills.

*Sanitation*—Spiders, as well as new and old webs, can be removed using a cobweb duster such as the Webster™ or a broom. A vacuum may be useful in situations with larger numbers of spiders.

*Exclusion*—Any crack in the exterior of a building through which spiders could enter should be sealed. Doors should have a tight-fitting weatherstrip on the bottom.

*Lighting*—Any steps

Figure 4: The webs of both spiders are very messy and are often very conspicuous around door and window frames.

taken to change exterior lighting away from white to yellow bulbs attracts far fewer of the flying insects used by these spiders and others as food is highly beneficial. Home fixtures can be equipped with yellow "bug" light bulbs while commercial lamps may require switching to sodium vapor bulbs.

*Insecticide Applications*—Insecticide applications may be needed where meshweb and flatweb weavers are in larger numbers. Monitoring traps placed inside near doorways may help capture spiders as they enter.

*Directed Contact Treatments*—Treatment of live, exposed spiders can be accomplished by directed contact treatment using a nonresidual aerosol insecticide.

*Dust Applications*—Application of a residual dust product into exterior cracks and holes as well as weep holes may be beneficial. Sealing of such cracks after application can be helpful with minimizing spider and pest invasions.

*Spot Treatments*— Spot treatments sites where spiders have constructed webs can be applied after webs have been removed. Sites include around windows and doors, around downspouts, and around exterior light fixtures. A WP, SC, or CS formulation should work well.

# RELATIVES OF SPIDERS

Within the Class Arachnida, there are several Orders of creatures. To be considered an arachnid, an animal must have mouthparts called chelicerae and eight legs when mature. The spiders are in the Order Araneae. Other Orders include the scorpions, mites and ticks, harvestman or daddylonglegs, vinegaroons, and a few others, which are somewhat rare. A few of the very rare arachnids that the pest professional would never encounter are purposefully left out.

**Mites and Ticks (Order Acari).** Mites and ticks are, by far, the most important group of arachnids. They have about 50,000 described species throughout the world with an estimate of maybe a million species overall. The ticks (Figure 1) are well known for being of supreme medical and veterinary importance for the bacterial and viral infections that they vector, sometimes resulting in severe morbidity and mortality in humans, their companion pets, and commercial livestock. Mites are severe pests from a medical standpoint as well as an agricultural angle where they cause massive destructive to crops and plants that are important to human society.

Figure 1: Ticks found around homes are typically associated with wildlife and carried inside by pets.

The mites and ticks will not be discussed further here because they would swamp the other arachnid Orders as well as the spiders in volume and significance. There is much written about these arachnids elsewhere such that in the small space alloted to non-spider arachnids here, that proper justice could not be done to their diversity and critical medical importance. The interested reader should seek other sources for more intense coverage.

**Harvestmen (Order Opiliones).** Harvestmen are also known as daddylonglegs, granddaddylonglegs, and phalangids. Unlike spiders, they have one body part that looks like a little pill with eight long, stilt-like legs with flexible feet (Figure 2, page 226). They usually have two eyes on a raised turret on the top of the body. The American species are pretty drab, being mostly tans and browns, but the harvestman in the tropics have brightly contrasting spots on their bodies and legs and are quite garish in appearance.

Harvestmen are usually found in damp situations, under rocks, logs, boards, and tarps, although some live out in the desert. They will shed a leg if it is grabbed as a way of escaping from a predator. They are typically solitary when they are active, however, when they overwinter in a place like a cave, thousands of them may amass such that, at first, it looks like a fungus or something else growing on the cave wall. Closer examination will reveal hundreds of little pill-like bodies and many more thousands of legs intertwined on the wall. Such gatherings can also be seen in crawlspaces and on buildings (Figure 3).

Figure 2: Harvestmen have one body region with eight long stilt-like legs and two eyes.

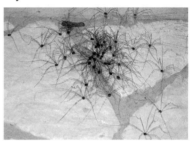

Figure 3: Harvestmen sometimes gather in large numbers in protected sites such as under the eave of this home.

A very well disseminated myth is that "daddylonglegs are the most poisonous spider in the world, but it has fangs that are too short to bite." Venom toxicologists have no idea where this myth started. In regard to harvestmen, first of all, they aren't spiders, and secondly, they lack venom glands and fangs. They do have little pincer-like chelicerae and people have received little nips from them but they do not have venom and, therefore, cannot be part of this myth in any way, shape, or form.

Inside, the most common place harvestmen will be found is damp crawlspaces and basements. Treatment of the areas where they might rest with residual liquid insecticides can be done, but in most cases, the numbers found inside will be few and control measures are generally not necessary.

**Scorpions (Order Scorpiones).** Scorpions are one of the oldest groups of arachnid, dating back millions of years (Figure 4, page 227). Their ancestors were found in the ocean, were easily recognizable as scorpions and reached lengths of five feet. (Imagine finding one of them in a crawlspace!) They are characterized by large claws with one rigid and one movable finger. They have a segmented abdomen and a segmented tail that ends in a bulb and sting (called the telson) (Figure 5). Scorpions are venomous and use their posterior sting to introduce venom under the cuticle or skin of its intended target.

Figure 4: Scorpions occur across the southern U.S. and as far north as the Ozark Mountains in Missouri.

Compared to other arachnids, scorpions have a long lifespans, living up to 25 or 30 years. They also can slow down their metabolisms during the inactive periods such that they do not require much food for survival and can go long periods between meals. Also, rare among arthropods, scorpions give birth to live babies, which crawl upon the mother's back (Figure 6, page 228). Depending on which family the scorpions belong to, the babies can either be randomly oriented on the mother's back or they may all align themselves in rows facing forward in the same direction, which is somewhat of a comical sight.

When feeding, scorpions use their claws to grab and eat prey. Some scorpions, like *Pandinus pandinus*, have massive claws and others, like the bark scorpion, *Centruroides sculpturatus*, have very fine claws. Typically a scorpion will not use its sting to subdue prey unless the prey is difficult to wrangle down. It does not want to waste its venom if it can subdue prey by just using physical means.

Figure 5: The sting, called the telson, at the tip of a scorpion's tail delivers venom to capture prey and provides self-defense.

Defensively, the venom and sting of a scorpion is made to cause immediate pain in mammals so that the "predator" is informed right away that it might be better to pursue the vegetarian plate that night. (However, grasshopper mice in the southwestern deserts have learned to attack and chew off the sting first, then attack and eat the rest of the scorpion.) The venom disrupts the nerve's elec-

Figure 6: Scorpions birth their babies live which then crawl onto their mother's back.

trical balance causing pain receptors to fire. Most of the scorpions in the United States are mildly harmful where the sting will be about the same as a bee sting. There is one potentially deadly scorpion in the United States. It is one of the bark scorpions, *Centruroides sculpturatus*, which is found in Arizona, and has established itself in localized areas of Orange County in Southern California, brought back by people who were camping and vacationing at the Colorado River on the California-Arizona border. (Note: Within a very short few years, the bark scorpion was known as *C. sculpturatus*, then it was renamed *C. exilicauda*, then it was changed back to *C. sculpturatus*. This will cause confusion for those who are trying to use the proper scientific names). Worldwide, scorpions are much more dangerous than spiders. In Mexico, India, Israel, and the Middle East, scorpions are responsible for thousands of human deaths each year.

Scorpions are best found during nights with little moonlight by using a blacklight under which scorpions fluoresce a light green color; for decades there was no practical explanation for them to glow like this. However, research in 2010 showed that you can exhaust a scorpion's supply of the fluorescent chemical, at which point, the scorpion does not avoid light. Therefore, it is thought fluorescence in scorpions is related to their ability to detect and, thereby, avoid being active under moonlit conditions where they would be more visible and more susceptible to predation.

Scorpions are most commonly found in warm climates and in deserts, however, they can be found in mountains where winters are cold and the ground is routinely covered with snow. The majority of the American species are found in the west, but one species, *Centruroides vittatus*, is widespread and found throughout much of the southern United States from Texas to Tennessee.

*Inspecting for Scorpions*—The best time to inspect for scorpions outside is at night. When exposed to the light from a black light bulb, scorpions glow a fluorescent green color which is easily seen. Around structures, expect to find scorpions roaming on top of rocks, patios, landscape timbers, and possibly the foundation.

One technique to determine if scorpions are a significant problem around a building is to thoroughly soak an area near the foundation with water. This area is then covered by a piece of cardboard or plywood with a rock placed on top. The moisture and the shelter provided by the board are attractive to scorpions. Checking the area in the next several days may reveal scorpions taking refuge there. If numerous scorpions have taken refuge, then the potential for a larger infestation in the building is higher.

*Managing Scorpion Infestations*—Briefly, scorpion management involves first, the removal of potential harborages outside near the structure; second, the application of residual insecticides (usually a dust) into cracks and voids where scorpions might be living; third, the sealing of exterior cracks to prevent scorpions from entering; and last, treatment of harborages where scorpions are found outdoors.

**Windscorpions (Order Solifugae).** Windscorpions are known by a great number of names such as wind spiders, camel spiders, and even barber spiders in South Africa because of the myth that they will cut your hair off while you sleep. They are neither scorpions nor spiders; their own unique name is either solpugids or solifuges but these names are used more often by arachnologists. They have puffy, segmented abdomens that look something like the Michelin Man cartoon icon; they also have two sets of jaws with beak-like projections (Figure 7). When they eat, the two halves of the head area move independently so that one half of the head can be moving up while the other is moving down. By doing this, they can mash and shear their prey into edible pieces. Quite frankly, there isn't an arachnid much uglier than a windscorpion when it is eating although tailless whipscorpions will give them a run for the money.

Figure 7: Windscorpions are known by several nicknames including wind spiders and camel spiders but they are neither spiders nor scorpions.

Windscorpions move very fast and typically do not rest easily. They are always in motion. If placed in a terrarium with sand, they might be seen bulldozing the sand around. Windscorpions do not have venom (except for one species in India), but they are very quick to defend themselves with a bite. These creatures are tenacious and fearless (and possibly stupid because they will launch an attack at a much larger creature). Because they are constantly in motion, it is really difficult to keep them in captivity. Compared to the other arachnids, very little is known about the biology of windscorpions merely because keeping them alive in captivity is so difficult. If you run across a live one, it would be best to turn it loose because its chances of survival in a plastic box are minimal.

Windscorpions are most commonly found in the hot, arid areas of the southwestern United States, however, they exist all the way up to North Dakota.

**Pseudoscorpions (Order Pseudoscorpiones).** These interesting and tiny creatures are not commonly found by the casual observer. They are typically very small (about 2 mm in body length) and usually found under rocks, in leaf litter, on tree bark. They look like little scorpions that have lost their tails and stings, having a rounded posterior abdomen instead (Figure 8). They do have venom glands in the tips of their pincers, however, it is merely for subduing small prey. Pseudoscorpions have no medical significance to humans.

Figure 8: Pseudoscorpions look like little scorpions without a tail.

One interesting aspect about pseudoscorpions is that they practice a dispersal mechanism called phoresy where they grab onto winged creatures (a bird, a beetle) and hide in their feathers or under their rigid wing coverings and travel around with their airborne carrier. Presumedly, they drop off later on, and possibly in an acceptable habitat.

Pseudoscorpions will rarely be seen inside buildings. It is possible for them to be carried inside in firewood. Occasionally, a sharp-eyed service technician or a homeowner will see one outside, collect it, and present it for identification.

**Tailed Whipscorpions (Order Uropygi).** The tailed whipscorpions are also known as vinegaroons. At the posteriormost portion of the abdomen, they have a long, thin tail, which they use like a whip to fling acetic acid at predators (Figure 9). (Acetic acid is a major component of vinegar, hence, the name vinegaroon.) They are able to bring this tail forward and aim it in the direction of a threat. Obviously, acetic acid in the eyes or on other sensitive tissue is an effective method of repelling something that is trying to eat you.

Figure 9: Tailed whipscorpions use their long tail to fling acetic acid at potential predators.

Tailed whipscorpions have large claws and a well-segmented abdomen. They are about 3 inches in length when mature. These creatures are found in Florida and Arizona.

**Tailless Whipscorpions (Order Amblypygi).** The tailless whipscorpions have massive front claws, armed with many spines (Figure 10). Their first pair of legs is modified such that they are extremely long and thin; they serve as antennae where they are swept out in front of the animal, tapping around similar to a blind person with a cane. If they sense a predator, they escape quickly.

Tailless whipscorpions hide under bark during the day and are active at night. Mature specimens can reach 2 inches in body length. They are found mostly in the tropics so in the United States, they exist almost exclusively in southern Florida.

Figure 10: In the tropics, tailless whipscorpions can be quite large but the species in the U.S. are much smaller.

# GLOSSARY

**Abdomen**—One of two major regions of a spider's body. Located at the rear of the spider, it contains the heart, silk glands, book lungs, and digestive and reproductive systems.

**Anterior**—Toward the front. For example, the anterior half of the abdomen is the front half where it joins the cephalothorax.

**Anterior Median Eyes**—The middle pair of eyes in the lower row of the eye arrangement; large and well-developed in the jumping spiders (Salticidae).

**Araneomorph**—A spider belonging to the true or common spiders of the Infraorder Araneomorphae. They possess diaxial (opposing) fangs. The other Infraorder is the Mygalomorphae which are the tarantulas and related species.

**Ballooning**—A method of dispersal employed by many spiderlings in which they are carried by air currents suspended from a length of silk.

**Book lungs**—The structure, consisting of a series of plates or leaves, used by spiders and scorpions for breathing.

**Calamistrum**—A comb of stiff hairs on the metatarsus of the fourth pair of legs of cribellate spiders. It is used to comb out multi-strand silk, which is exuded from the cribellum.

**Carapace**—The hardened shield that covers the cephalothorax. It protects the internal organs and acts as an anchor for the muscles of the sucking stomach.

**Cephalothorax**—The anterior or front region or section of the spider's body. Equivalent to the head and thorax of insects.

**Chelicerae**—The jaws of arachnids. Each half consists of a basal segment upon which is hinged the sharp fang which injects the venom.

**Claw Tufts**—The bunch of hairs at the tip of the tarsus in those spiders with only two claws.

**Clypeus**—The space between the anterior row of eyes and the anterior edge of the carapace.

**Copulatory Organ**—That portion of the distal (end) segment of the male palpus used in mating.

**Coxa**—The first leg segment attached to the body between the cephalothorax and the trochanter.

**Cribellate**— Spiders that possess a cribellum (see definition below). This applies to only a small number of species of the araneomorph spiders.

**Cribellum**—A special plate through which multi-stranded silk is secreted. It is derived from a pair of modified spinnerets. Found only in cribellate spiders.

**Diaxial**—Chelicerae of the araneomorph spiders where the fangs come together in opposing fashion like the opening and closing of pliers.

**Ecribellate**—Adjective applying to a spider not provided with a cribellum. Most of the true spiders are ecribellate.

**Endites**—Mouthparts on the ventral surface of the body in front of the sternum which spiders use during feeding

**Epigynum**—A sclerite associated with the reproductive openings in the female of most spiders. The opening on the underside of her abdomen into which the male places his palp during mating.

**Exoskeleton**—The skeleton of arthropods which is actually its outside skin.

**Fang**—The hardened, pointed structure at the end of the chelicerae which the spider uses for grabbing prey and injecting venom.

**Femur**—The third leg segment, between the trochanter and the patella.

**Haemolymph**—The blood of spiders.

**Hair Tufts**—Tufts of hair particularly on the legs, palps and face of spiders such as the jumping spiders. Often colored and set in specific patterns.

**Invertebrates**—Any animal that lacks a backbone (e.g., insects, spiders, worms, etc.).

**Latrodectism**—The medically significant symptoms expressed after a bite by a black widow spider (*Latrodectus* species).

**Loxoscelism**— The medically significant symptoms expressed after a bite by a recluse spider (*Loxosceles* species).

**Metatarsus**—The sixth leg segment from the body between the tibia and tarsus.

**Mygalomorphae**—Infraorder of spiders with paraxial fangs, 4 book lungs. Includes tarantulas.

**Molt**—Also known as ecdysis; a process of shedding the exoskeleton.

**Necrosis**—The death or decay of tissue, in this book, as a result of the bite of a spider.

**Oviduct**—The tube leading away from the ovary through which the eggs pass.

**Palp**—Abbreviated word for "pedipalp" (see "pedipalp" for definition).

**Paraxial**—Chelicerae of the Infraorder Mygalomorphae that project forward horizontally with the fang moving in a plane parallel to the body like the striking pattern of a rattlesnake.

**Patella**—The fourth leg segment, between the femur and tibia.

**Pedipalp**—The second pair of appendages of the cephalothorax, behind the chelicerae, but in front of the legs. In males, this is transformed into the mating organ. In females, it looks like a miniature leg.

**Posterior Median Eyes**—The middle pair of eyes in the back row. In wolf spiders, these are very large, prominent, and forward facing.

**Rastellum**—A specialized scraping structure on the chelicerae of some mygalomorphs which they use for digging.

**Sexual Dimorphism**—When males and females of the same species are of different sizes, general shape, or color.

**Spermatheca**—The sac-like structure in female spiders in which sperm from the male is received and often stored.

**Spiderlings**—Newly hatched spiders. Generally called this through the first few molts.

**Spinnerets**—Paired appendages at the posterior end of the abdomen below the anus, through which silk is exuded. Spiders have either two or three pairs.

**Spiracle**—The external opening of the tracheal system; a breathing pore.

**Stabilimentum**—A heavy band of silk placed in the web, usually at the center, by some orb-weaving spiders, mostly the garden spiders (*Argiope* species).

**Sternum**—The sclerite forming the ventral wall, or underside, of the cephalothorax.

**Sucking/Pumping Stomach**—The stomach of spiders that is capable of pumping digestive fluids over or into the body of the prey and then also sucking the fluids back into the stomach.

**Tarsus**—The seventh and last leg segment, after the metatarsus.

**Tergites**—A sclerotized or hardened area on the dorsal (top) side of the abdomen.

**Tibia**—The fifth segment of the leg, between the patella and the metatarsus.

**Tracheoles**—The very thin, terminal segments of the respiratory tubes on the inside of the abdomen.

**Trochanter**—The second leg segment, between the coxa and the femur. Very small and hard to see unless you bend the spider's leg up over the carapace.

**Urticating Hairs**—Hairs of mygalomorph spiders that are easily broken or rubbed off and which are irritating to the skin or nasal passages of people and animals.

**Venom**—A mixture of proteins designed to immobilize and kill prey. Mostly used for offensive purposes to secure a meal, although some larger spider species also use it to defend themselves.

# REFERENCES

Bennett, R.G. and R.S. Vetter. **2004**. An approach to spider bites: erroneous attribution of dermonecrotic lesions to brown recluse or hobo spider bites in Canada. Canadian Family Physician 50: 1098-1101.

Binford, G.J. **2001**. An analysis of geographic and intersexual chemical variation in venoms of the spider *Tegenaria agrestis* (Agelenidae). Toxicon. 39: 955-968.

Borror, D.J. and D.M. DeLong. **1992**. An Introduction to the Study of Insects, 6th ed., Saunders College Publishing, Philadelphia, Pa.

Brown, K.S., J.S. Necaise, and J. Goddard. **2008**. Additions to the known U.S. distribution of *Latrodectus geometricus* (Araneae: Theridiidae). J. Med. Entomol. 45: 959-962.

Bush, S.P., P. Giem, and R.S. Vetter. **2000**. Green lynx spider (*Peucetia viridans*) envenomation. Amer. J. Emerg. Medicine 18: 64-66.

Crawford, C. and D.K. Vest. **1989**. The hobo spider and other European house spiders. Univ. of Wash., Burke Museum Educational Bulletin No. 1, 4 pp.

Dominguez, T.J. **2004**. It's not a spider bite, it's community-acquired methicillin-resistant *Staphylococcus aureus*. J. Amer. Board Fam. Pract. 17: 220-226.

Ebeling, W. **1975**. Urban Entomology. Los Angeles: University of California.

Edwards, G.B. **1985**. The common house spider, *Achaearanea tepidariorum*. Fla. Dept. of Ag. & Consum. Svcs., Entomology Circular No. 279.

Edwards, G.B. **1988**. A shiny orb-weaver, *Gasteracantha cancriformis*, in Florida. Fla. Dept. of Ag. & Consum. Svcs., Entomology Circular No. 308.

Evans, D.L. and J.O. Schmidt. **1990**. Insect Defenses. State University of New York Press, Albany, N.Y.

Fink, L.S. **1984**. Venom-spitting by the green lynx spider, *Peucetia viridans* (Araneae, Oxyopidae). J. Arachnol. 12: 372-373.

Foelix, R.F. **2010**. Biology of Spiders. 3$^{rd}$ ed. Oxford University Press, Cambridge, Mass.

Frithsen, I.L., R.S. Vetter and I.C. Stocks. **2007**. Reports of envenomation by brown recluse spiders exceed verified specimens of *Loxosceles* spiders in South Carolina. J. Amer. Board Fam. Med. 20: 483-488.

Gaver-Wainwright, M. M., R. S. Zack, M. J. Foradori, and L. C. Lavine. **2011**. Misdiagnosis of spider bites: bacterial associates, mechanical pathogen transfer, and hemolytic potential of venom from the hobo spider, *Tegenaria agrestis* (Araneae: Agelenidae). J. Medical Entomol. 48: 382-388.

Gertsch, W.J. **1979**. American Spiders. Van Nostrand Reinhold, New York, N.Y.

Goddard, J., S. Upshaw, D. Held, and K. Johnson. **2008**. Severe reaction from envenomation by the brown widow spider, *Latrodectus geometricus* (Araneae: Theridiidae). Southern Med. J. 101: 1269-1270.

Greene, A., N.L. Breitsch, T. Boardman, B.B. Pagac, E. Kunickas, R.K. Howes, and P.V. Brown. **2009**. The Mediterranean recluse spider, *Loxosceles rufescens* (Dufour): an abundant but cryptic inhabitant of deep infrastructure in Washington, D.C. area (Arachnida: Araneae: Sicariidae). Amer. Entomol. 55: 158-167.

Greene, A., J. A. Coddington, N.L. Breisch, D.M. De Roche, and B.B. Pagac, Jr. **2010**. An immense concentration of orb-weaving spiders with communal webbing in a man-made structural habitat. Amer. Entomol. 56: 146-156.

Hedges, S. **1988**. The Remarkable World of Spiders. Pest Control Technology. 16(10): 32-34, 36, 113.

Hedges, S. **2000**. Profitable Spiders. Pest Control Technology. 28(9): 40-42, 43.

Hedges, S. **2003**. Spider-Man. Pest Control Technology. 31(8): 64, 65, 68, 73.

Hedges, S. and R. Vetter. **2001**. Got Recluses? Pest Control Technology. 29(3): 46-53.

Hillyard, P. **1994.** The Book of the Spider: From Arachnophobia to the Love of Spiders. Random House. New York, 218 pp.

Hinkle, N.C. **2000**. Delusory parasitosis. Amer. Entomol. 46: 17-25.

Hite, J.M., W.J. Gladney, J.L. Lancaster and W.H. Whitcomb. **1966**. Biology of the brown recluse spider. Univ. of Arkansas Agric. Exper. Sta. Bulletin #711, 26 pp.

Hogue, C.L. **1993**. Insects of the Los Angeles Basin, 2nd Ed. Natural History Museum of Los Angeles County, Los Angeles, Calif.

Holper, J. 2007. What can brown do for you? Pest Control Technology 35: 87-90.

Kaston, B.J. **1970**. Comparative biology of American black widow spiders. Trans. San Diego Soc. Nat. Hist. 16: 33-82.

Kaston, B.J. **1981**. Spiders of Connecticut (revised edition). Bull. Connecticut Geol. Nat. Hist. Survey 70: 1020 pp.

Kloock, C. T., A. Kubli, and R. Reynolds. **2010**. Ultraviolet light detection: a function of scorpion fluorescence. J. Arachnol. 38: 441-445.

Levi, H.W. and L.R. Levi. **1990**. Spiders and Their Kin, Golden Guide. Western Pub. Co. Inc., Racine, Wis.

Marshall, S.D. **1996**. Tarantulas and other arachnids: a complete pet owner's manual. Barron's Educational Series, Hauppauge, N.Y.

Mullen, G. R., and R. S. Vetter. **2009**. Spiders (Araneae). Ch. 24, pp. 403-424. *In:* Medical and Veterinary Entomology, 2$^{nd}$ edition, G.R. Mullen, L.A. Durden (eds.), Elsevier Press.

Müller, G.J. **1993**. Black and brown widow spider bites in South Africa: a series of 45 cases. African Med. J. 83: 399-405.

Platnick, N.I. **1997**. Two reports of envenomation by the spider *Trachelas tranquillus* (Hentz). Cincinnati J. Med 52: 194.

Preston-Mafham, R. **1991**. The Book of Spiders and Scorpions. Cresent Books, New York, N.Y.

Preston-Mafham, R.&K. **1984**. Spiders of the World. Facts On File Publications, New York, N.Y.

Ramires, E.N., A.V.L. Retzlaff, L.R. Deconto, J.D. Fontana, F.A. Marques, and E. Marques-Da-Silva. **2007**. Evaluation of the efficacy of vacuum cleaners for the integrated control of brown spider *Loxosceles intermedia*. J. Venom Anim. Toxins 13: 607-619.

Russell, F.E. **1970**. Bite of the spider *Phidippus formosus*: case history. Toxicon 8: 193-194.

Sandidge, J. and J. L. Hopwood. **2005**. Brown recluse spiders: A review of biology, life history and pest management. Trans. Kansas Acad. Sci. 108: 99-108.

Sarno, P.A. **1973**. A new species of *Atypus* (Araneae: Atypidae) from Pennsylvania. Entomol. News 84: 37-51.

Schenone, H., A. Rojas, H. Reyes, F. Villaroel, and G. Suarez. **1970.** Prevalence of *Loxosceles laeta* in houses in central Chile. Amer. J. Trop. Med. Hyg. 19: 564-567.

Stropa, A.A. **2010**. Effect of architectural angularity on refugia selection by the brown spider, *Loxosceles gaucho*. Med. Vet. Entomol. 24: 273-277.

Swanson, D.L. and R.S. Vetter. **2005**. Bites of brown recluse spiders and suspected necrotic arachnidism. New Engl. J. Med. 352: 700-707.

Turnbull, A.L. **1960**. Ecology of the true spider (Araneomorphae). Ann.

Rev. Entomol. 18: 305-348.

Ubick, D., P. Paquin, P.E. Cushing, and V. Roth. (eds.) **2005**. Spiders of North America: an identification manual. American Arachnological Society.

Uetz, G.W. **1974**. Envenomation by the spider *Trachelas tranquillus* (Araneae: Clubionidae). J. Med. Entomol. 10: 227.

Vetter, R.S. **1998**. Envenomation by an agelenid spider, *Agelenopsis aperta*, previously considered harmless. Ann. Emerg. Med. 32: 739-741.

Vetter, R.S. **2005**. Arachnids submitted as suspected brown recluse spiders (Araneae: Sicariidae): *Loxosceles* species are virtually restricted to their known distributions but are perceived to exist throughout the United States. J. Med. Entomol. 42: 512-521.

Vetter, R. **2005**. [They're not] spider bites. Pest Control Technology 33(4): 86-88.

Vetter, R.S. **2006.** Hobo spider. Univ. Calif. Pest Notes #7488, 3pp. http://www.ipm.ucdavis.edu/PMG/PESTNOTES/pn7488.html

Vetter, R.S. **2008**. Spiders of the genus *Loxosceles* (Araneae: Sicariidae): a review of biological, medical and psychological aspects regarding envenomations. J. Arachnol. 36: 150-163.

Vetter, R.S. **2009**. Arachnids misidentified as brown recluse spiders by medical personnel and other authorities in North America. Toxicon 54: 545-547.

Vetter, R.S. **2009**. The distribution of the brown recluse spider in the southeastern quadrant of the United States in relation to loxoscelism diagnoses. Southern Med. J. 102: 518-522.

Vetter, R.S. **2010**. Myths based in science and medicine – how they initiate, propagate, and the role of peer-review research in dispelling them. Perspect. Agric. Veterin. Sci. Nutrition Natur. Resources 5, #041, 7 pp.

Vetter, R. **2010**. Bite or bacteria? Pest Management Professional 78(1): 42-46.

Vetter, R.S. **2011**. Scavenging in spiders (Araneae) and its relationship to the pest management of the brown recluse spider. J. Econ. Entomol. 104: 986-989.

Vetter, R.S. **2011**. Spiders. *In:* Handbook of Pest Control, 10th Ed. S.A. Hedges, ed. Mallis Handbook & Technical Training Company, Cleveland, Ohio. pp. 1082-1117.

Vetter, R.S. and D.K. Barger. **2002**. An infestation of 2,055 brown recluse spiders (Araneae: Sicariidae) and no envenomations in a Kansas home: implications for bite diagnoses in nonendemic areas.

J. Med. Entomol. 39: 948-951.

Vetter, R.S. and S.P. Bush. **2002**. Chemical burn misdiagnosed as brown recluse spider bite. Amer. J. Emerg. Medicine 20: 68-69.

Vetter, R.S. and S.P. Bush. **2002**. Reports of presumptive brown recluse spider bites reinforce improbable diagnosis in regions of North America where the spider is not endemic. Clinical Infectious Diseases. 35: 442-445.

Vetter, R.S. and S. Hillebrecht. **2008**. On distinguishing two often-misidentified genera (*Cupiennius*, *Phoneutria*) (Araneae: Ctenidae) of large spiders found in Central and South American cargo shipments. Amer. Entomol. 54: 82-87.

Vetter, R.S. and G.K. Isbister. **2004**. Do hobo spider bites cause dermonecrotic injuries? Ann. Emerg. Medicine. 44: 605-607.

Vetter, R.S. and G.K. Isbister. **2006**. Verified bites by the woodlouse spider, *Dysdera crocata*. Toxicon. 47: 826-829.

Vetter, R.S. and G.K. Isbister. **2008**. Medical aspects of spider bites. Ann. Rev. Entomol. 53: 409-429.

Vetter, R.S. and M.K. Rust. **2008**. Refugia preferences by the spiders *Loxosceles reclusa* and *Loxosceles laeta* (Araneae: Sicariidae). J. Med. Entomol. 45: 36-41.

Vetter, R.S. and M.K. Rust. **2010**. Influence of spider silk on the refugia preferences of the recluse spiders *Loxosceles reclusa* and *Loxosceles laeta* (Araneae: Sicariidae). J. Econ. Entomol. 103: 808-815.

Vetter, R.S. and M.K. Rust. **2010**. Periodicity of molting and resumption of post-molt feeding in the brown recluse spider, *Loxosceles reclusa* (Araneae, Sicariidae). J. Kansas Entomol. Soc. 83: 306-312.

Vetter, R.S. and P.K. Visscher. **2000**. Oh, what a tangled web we weave: the anatomy of an internet spider hoax. Amer. Entomol. 46: 221-223.

Vetter, R.S., P.E. Cushing, R.L. Crawford, and L.A. Royce. **2003**. Diagnoses of brown recluse spider bites (loxoscelism) greatly outnumber actual verifications of the spider in four western American states. Toxicon. 42: 413-418.

Vetter, R.S., G.B. Edwards, and L.F. James. **2004**. Reports of envenomation by brown recluse spiders (Araneae: Sicariidae) outnumber verifications of *Loxosceles* spiders in Florida. J. Med. Entomol. 41: 593-597.

Vetter, R.S., C.P. Flanders, and M.K. Rust. **2009**. The ability of spiderlings of a widow spider, *Latrodectus hesperus* (Araneae: Theridiidae) to pass through different size mesh screen: implications for exclusion from air intake ducts and greenhouses. J. Econ. Entomol. 102: 1396-1398.

Vetter, R.S., N.C. Hinkle and L.M. Ames. **2009**. Distribution of the brown recluse spider (Araneae: Sicariidae) in Georgia with a comparison of poison center reports of envenomations. J. Med. Entomol. 46: 15-20.

Vetter, R.S., G.K. Isbister, S.P. Bush, and L.J. Boutin. **2006**. Verified bites by *Cheiracanthium* spiders in the United States and Australia: where is the necrosis? Amer. J. Trop. Med. Hyg. 74: 1043-1048.

Vetter, R.S., B.B. Pagac, R.W. Reiland, D.T. Bolesh, and D.L. Swanson. **2006**. Skin lesions in barracks: consider community-acquired methicillin-resistant *Staphylococcus aureus* infection instead of spider bites. Military Medicine. 171: 830-832.

Vetter, R.S., D.A. Reierson, and M.K. Rust. **2011**. Cobweb management and control of the cellar spider, *Holocnemus pluchei* (Araneae: Pholcidae) in the eaves of buildings. J. Econ. Entomol. 104: 601-606.

Vetter, R.S., A.H. Roe, R.G. Bennett, C.R. Baird, L.A. Royce, W.T. Lanier, A.L. Antonelli, and P.E. Cushing. **2003**. Distribution of the medically-implicated hobo spider (Araneae: Agelenidae) and its harmless congener, *Tegenaria duellica,* in the United States and Canada. J. Med. Entomol. 40: 159-164.

Vincent, L.S., R.S. Vetter, W.R. Wrenn, J.K. Kempf, and J.E. Berrian. **2008**. The brown widow spider, *Latrodectus geometricus* C. L. Koch, 1841 in southern California. Pan-Pac. Entomol. 84: 344-349.

Vollrath, F. **1992**. Spider webs and silk. Scientific American. 266 (3): 70-76.

# ILLUSTRATION/ PHOTO CREDITS

| Page | Source |
|---|---|
| 15 | Mark. S. Lacey |
| 16 | Richard S. Vetter |
| 17 | Richard S. Vetter |
| 18 | From How to Know the Spiders |
| 21 | Richard S. Vetter |
| 22 | Richard S. Vetter, Terminix |
| 23 | Richard S. Vetter |
| 24 | Richard S. Vetter |
| 25 | Richard S. Vetter |
| 26 | Richard S. Vetter |
| 28 | Richard S. Vetter |
| 29 | Richard S. Vetter |
| 30 | Stoy A. Hedges, Richard S. Vetter |
| 31 | Stoy A. Hedges |
| 33 | Mark S. Lacey |
| 34 | Richard S. Vetter |
| 35 | Richard S. Vetter |
| 36 | Stoy A. Hedges |
| 37 | Richard S. Vetter |
| 39 | Vance Walker |
| 40 | Mark S. Lacey |
| 42 | Richard S. Vetter |
| 43 | Richard S. Vetter |
| 44 | Stoy A. Hedges, Richard S. Vetter |
| 46 | Richard S. Vetter |
| 48 | Richard S. Vetter |
| 49 | Richard S. Vetter |
| 51 | Richard S. Vetter |
| 53 | Richard S. Vetter |
| 54 | Richard S. Vetter |
| 55 | Stoy A. Hedges |
| 57 | Stoy A. Hedges |
| 58 | Richard S. Vetter |

| | |
|---|---|
| 60 | Stoy A. Hedges |
| 61 | Richard S. Vetter |
| 62 | Richard S. Vetter |
| 65 | Stoy A. Hedges |
| 66 | Janie Stvan |
| 68 | Richard S. Vetter, Janie Stvan |
| 69 | Richard S. Vetter |
| 70 | Richard S. Vetter |
| 71 | Richard S. Vetter, Janie Stvan |
| 72 | Janie Stvan, Richard S. Vetter |
| 73 | Richard S. Vetter |
| 74 | Janie Stvan, Richard S. Vetter |
| 78 | Stoy A. Hedges |
| 81 | Janie Stvan |
| 82 | Stoy A. Hedges |
| 83 | Paul G. Curtis |
| 86 | Stoy A. Hedges |
| 88 | Stoy A. Hedges |
| 89 | Stoy A. Hedges |
| 103 | Richard S. Vetter |
| 104 | Janie Stvan |
| 105 | Richard S. Vetter |
| 107 | Stoy A. Hedges |
| 109 | Premaphotos Wildlife |
| 110 | Janie Stvan |
| 111 | Premaphotos Wildlife |
| 113 | Richard S. Vetter |
| 115 | Paul G. Curtis |
| 116 | Richard S. Vetter |
| 119 | Paul G. Curtis, Richard S. Vetter |
| 120 | Richard S. Vetter |
| 121 | Richard S. Vetter, Stoy S. Hedges |
| 123 | Janie Stvan |
| 124 | Richard S. Vetter |
| 129 | Janie Stvan, Richard S. Vetter |
| 130 | Stoy A. Hedges, Richard S. Vetter |
| 135 | Janie Stvan, Richard S. Vetter |
| 136 | From How to Know the Spiders, Richard S. Vetter |
| 139 | Paul G. Curtis, Richard S. Vetter |

| Page | Credit |
|---|---|
| 140 | Richard S. Vetter |
| 141 | Stoy A. Hedges |
| 143 | Janie Stvan |
| 144 | Richard S. Vetter |
| 147 | Janie Stvan |
| 148 | Janie Stvan |
| 150 | Richard S. Vetter |
| 153 | Paul G. Curtis, Richard S. Vetter |
| 154 | Richard S. Vetter |
| 155 | Richard S. Vetter |
| 157 | Paul G. Curtis, Richard S. Vetter |
| 158 | Richard S. Vetter |
| 161 | Janie Stvan, Richard S. Vetter |
| 162 | Richard S. Vetter |
| 165 | Paul G. Curtis, Richard S. Vetter |
| 166 | Richard S. Vetter |
| 167 | Richard S. Vetter |
| 168 | Richard S. Vetter |
| 169 | Richard S. Vetter |
| 172 | Richard S. Vetter |
| 176 | Richard S. Vetter |
| 177 | Stoy A. Hedges |
| 178 | Richard S. Vetter |
| 179 | Stoy A. Hedges |
| 183 | Janie Stvan, Richard S. Vetter |
| 184 | Richard S. Vetter |
| 185 | Richard S. Vetter, Janie Stvan |
| 186 | Stoy A. Hedges, Richard S. Vetter |
| 187 | Richard S. Vetter |
| 188 | Al Greene |
| 189 | Stoy A. Hedges |
| 191 | From How to Know the Spiders, Richard S. Vetter |
| 192 | Richard S. Vetter |
| 193 | Richard S. Vetter |
| 194 | Marco Metzger |
| 197 | Janie Stvan, Richard S. Vetter |
| 198 | Richard S. Vetter |
| 199 | Richard S. Vetter |
| 200 | Richard S. Vetter |

| | |
|---|---|
| 201 | Stoy A. Hedges |
| 205 | From Emerton, Richard S. Vetter |
| 206 | Richard S. Vetter |
| 207 | Janie Stvan |
| 211 | From How to Know the Spiders |
| 212 | Richard S. Vetter |
| 213 | Richard S. Vetter |
| 214 | Richard S. Vetter |
| 215 | Richard S. Vetter |
| 217 | Richard S. Vetter |
| 218 | Richard S. Vetter |
| 221 | Paul G. Curtis |
| 222 | Richard S. Vetter |
| 223 | Richard S. Vetter |
| 225 | Richard S. Vetter |
| 226 | Stoy A. Hedges |
| 227 | Stoy A. Hedges, Richard S. Vetter |
| 228 | Stoy A. Hedges |
| 229 | Richard S. Vetter |
| 230 | Richard S. Vetter |
| 231 | Richard S. Vetter |

# THE PCT TECHNICAL RESOURCE LIBRARY

## *New Releases*

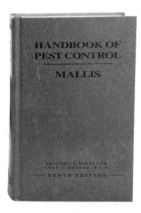

### Handbook of Pest Control, 10th Edition

*By Arnold Mallis; Stoy A. Hedges, Editorial Director*

The Mallis Handbook of Pest Control is the leading reference book in the pest management industry, providing a comprehensive overview of the biology, behavior, and control of hundreds of common and not-so-common structural pests. The 1,600-page publication provides pest control operators with the information needed to deliver effective, environmentally conscious pest management services.

Written in easy-to-understand language, the Mallis Handbook of Pest Control features more than 1,000 photographs and insect illustrations, including comprehensive insect keys and a special 70-page, color photo identification section.

For detailed information about the Mallis Handbook of Pest Control, visit www.mallishandbook.com.

# New Releases

## The Service Technician's Field Manual

*By William H Robinson, Ph.D.*

The Service Technician's Field Manual is packed with valuable information about situation-specific pest control. This 218-page, fully-illustrated manual is perfect for on-the-job reference, in-class technical training, or preparing for state certification exams. Topics include: IPM, non-chemical control methods, chemical insecticides and formulations, termite and wood-infesting insect control, structural fumigation, application equipment, and rodent control.

# *New Releases*

## PCT Technician's Handbook, 4th Edition

*By Dr. Richard Kramer*

The PCT Technician's Handbook is a guide to the identification and management of numerous pests including ants, blood feeders, cockroaches, flies, wood-destroying insects and vertebrate pests. It contains two-page pest profiles that include a description and illustration of the pest, along with its biology, habits and control.

**For pricing and ordering information, call the
PCT Bookstore at 800.456.0707 or visit the online store
at www.pctonline.com/store.**

# PCT FIELD GUIDES

## PCT Field Guide for the Management of Structure-Infesting Ants, 3rd Edition
*By Stoy A. Hedges*

Ants are the industry's #1 pest, representing millions of dollars in annual revenue. Unique behavioral characteristics, rapidly changing dietary needs, and dozens of structure-infesting species all make control difficult. Where can PCOs find the most practical, up-to-date information on this most challenging of all pests? The answer is simple: The PCT Field Guide for the Management of Structure-Infesting Ants, 3rd Edition, covers basic ant biology, inspection tips, and treatment strategies. The pocket size guide contains an ant identification guide, taxonomic key, and a full-color photo identification section.

# PCT Field Guide for the Management of Structure-Infesting Flies

*By Stoy A. Hedges*

A comprehensive publication featuring in-depth profiles of the three classes of flies frequently encountered by PCOs: small flies, filth flies, and nuisance flies. The book includes identification tips, a brief taxonomic key, practical management strategies, case histories, and a listing of leading fly control product manufacturers. The guide also offers a special color photograph section and the individual pest profiles feature key biology, physical, and behavioral characteristics.

## PCT Field Guide for the Management of Structure-Infesting Beetles, Volume I and II

*By Stoy A. Hedges and Dr. Mark S. Lacey*

These handy, pocket size guides to the biology, behavior, and control of beetles are written in easy-to-understand language and feature information on key identifying biological, behavioral, and physical characteristics of hundreds of structure-infesting beetles. Volume I covers Hide and Carpet and Wood-Boring Beetles, while Volume II features Stored Pest and Overwintering Beetles. Both volumes contain hundreds of detailed illustrations and a color identification section.

**For pricing and ordering information, call the PCT Bookstore at 800.456.0707 or visit the online store at www.pctonline.com/store.**